ENERGY SCIENCE, ENGINEERING AND T

BIOENERGY SYSTEMS, BIOLOGICAL SOURCES AND ENVIRONMENTAL IMPACT

Energy Science, Engineering and Technology

Additional books in this series can be found on Nova's website
under the Series tab.

Additional e-books in this series can be found on Nova's website
under the e-book tab.

Environmental Science, Engineering and Technology

Additional books in this series can be found on Nova's website
under the Series tab.

Additional e-books in this series can be found on Nova's website
under the e-book tab.

ENERGY SCIENCE, ENGINEERING AND TECHNOLOGY

BIOENERGY SYSTEMS, BIOLOGICAL SOURCES AND ENVIRONMENTAL IMPACT

MICHEL C. ALLARD
EDITOR

nova publishers
New York

For permission to use material from this book please contact us:
Telephone 631-231-7269; Fax 631-231-8175
Web Site: http://www.novapublishers.com

Library of Congress Cataloging-in-Publication Data

ISBN: 978-1-62417-330-1

Published by Nova Science Publishers, Inc. † New York

CONTENTS

PREFACE

In this book, the authors present current research in the study of bioenergy systems and their biological sources and environmental impact. Topics discussed include the assessment of sustainability of bioenergy systems; soil loss as a forgotten aspect of biofuel production; effective biocatalysts for lignocellulosic biofuel by a proteomics study of fungal secretome; enhancing the interface between agricultural production, and tree biomass/bioenergy systems to improve farm-scale productivity in Western Australia; hydrogen and ethanol comparison as fuels for internal combustion engines; and hydrogen production by mixed cultures.

Chapter 1 – The use of biomass -term includes materials of organic origin unfossilized- is being promoted worldwide mainly for two reasons: i) its energy contribution and ii) involvement in reducing CO_2 emissions. Unlike other renewable energy sources, the biomass includes a variety of resources all of organic nature but with different characteristics. Bioenergy systems based on biomass are complex because they cover various stakeholders and economic sectors, different legal frameworks, resource types, implementation options and energy conversion processes, which generate systems with particular characteristics that are included in certain social and environmental contexts. The fact that the biomass is associated with the benefits mentioned and even many other benefits (partial replacement of fossil fuels, regional and local availability, source of carbon neutral energy, low or no cost, versatility for generating electricity, heat, solid or gas or liquid fuels) not imply that their use is inherently sustainable. The condition of sustainability - which does not indicate a fixed target but a path to a more desirable stage - depends on how a group perceives the reality and what their interests are in a given historical and temporal moment. Therefore this concept must be defined in each particular context. There should also be defined criteria and indicators to monitor the trend of the bioenergy system towards higher or lower levels the sustainability in the context analyzed. In the Lerma Valley, province of Salta, Argentina, where there are different resources of biomass, four bioenergy systems (energy systems based on biomass) were defined for analysis. By means of a Multi-Criteria Assessment, the four systems were evaluated in issues identified and agreed by means of a survey of international experts. Three of the energy systems on the basis of biomass show a level of sustainability between medium and high (50 to 75% of a perfect system). Though no bioenergy system reaches 100% of sustainability, they present valuables qualities for its implementation. Analyses identify areas with greater shortcoming and recognition of the dimensions that require the application of policies or corrective tools. The simultaneous and planned application of the proposed bioenergy systems could have a positive impact on local sustainability of the Lerma Valley.

Chapter 2 – The increasing demand for energy and the decreasing reserves of fossil fuels have turned the focus of governments to the production of energy from biomass (biofuels, biogas). Since biomass is currently the only renewable source of fixed carbon it is an ideal source for the production of transport fuels. However, because of the high production targets set by various governments, much fertile agricultural land is diverted from food to bioenergy production. Moreover, because of agricultural practices, sol is lost or degraded. Soil is one of the most important but equally neglected resources. It not only produces food/forage/bioenergy but also sequesters greenhouse gases, filters rain water before reaching the groundwater and supports biodiversity. It is important to select suitable bioenergy plants, such as perennial grasses and short rotation wood coppice instead of row crops such as corn. Furthermore, attention should be given to no tillage practice with an incorporation of crops residues into the soils, thereby increasing the sustainability of biofuel production. In addition, soils which are already degraded could be used as an alternative to fertile land. Attention should be given to not only conserving land, but to improving and perhaps even restoring contaminated and degraded soils in the long term.

Chapter 3 – Due to the impact of energy crisis and global warming on socioeconomic structure, clean renewable energy is urgently required to replace the fossil energy. Lignocellulosic biomass is the most abundant renewable bioresource that can contribute to significant amount of global energy demand to alleviate global warming. To convert solid biomass to higher value biofuel, the current bottleneck is the hydrolysis of the biomass to monomeric sugar for further fermentation to biofuel. Lignocellulosic biomass is multifarious polymer composed of a complex mixture of cellulose, hemicellulose, and lignin; and its bioconversion to monomeric sugars requires a large number of enzymes. Fungi produce an arsenal of extracellular enzymes, the secretome, that catalyses degradation of complex lignocellulosic biomass. To accelerate the biotechnological application of these enzymes in lignocellulosic biofuel, more recently, proteomics technology has been applied to explore the comprehensive secretory fungal lignocellulolytic enzymes during lignocellulosic biomass degradation. This chapter illustrates the application of state-of-the-art proteomics technology for fungal proteomics, particularly emphasizes on secretory proteome (secretome) and also highlights new insights into the diversity of expressed fungal lignocellulolytic enzymes during cellulose and lignocellulose degradation. A broad understanding of the quantitative composition of extracellular secreted lignocellulolytic enzymes and the underlying synergetic mechanism may shed light on the fundamental understanding and further its application to lignocellulosic biofuel and sustainable development. The lignocellulosic biomass composition, their type, nature and complexity has been correlated with fungal lignocellulolytic enzymes identified by mass spectrometry.

Chapter 4 – Brazil is in the list of the largest biofuels producers in the world. The country presents biodiversity and environmental characteristics that make it suitable for the cultivation of several biomasses with high productivity, such as sugarcane (*Saccharum* spp.) and soybean (*Glycine max*), which are used for the production of bioethanol and biodiesel, respectively. This chapter discusses cultivation aspects, geographical distribution, production costs and the historical series of the production of the main biomasses cultivated in Brazil, as well as comments on the other agricultural resources that are being studied for the production of biofuels and green chemicals, such as castor fruit (*Ricinus communis*) and oil palm fruit (*Elaeis guineensis*). The current and potential uses of these materials in the context of sustainable products and biorefineries are also discussed.

Chapter 5 – Shifting government policies and degrading terms of trade are leading to a range of tree biomass related options competing with food production on agricultural lands, including bioenergy, forestry, conservation legislation, and carbon biosequestration markets. Tree bioenergy plantations need not be monocultural and homogenous, and can be successfully incorporated into existing agro-ecological systems to increase primary productivity and food security while providing cost-effective bioenergy resources. This work examines tree integration in relation to wind speed, turbulence, humidity, evaporation, transpiration, temperature, water competition, solar use efficiency, frost, erosion, and fodder. These effects are examined in terms of farm-scale limiting factors in agro-ecological systems (water, sunlight, wind, frosts, fodder, etc.), alongside research data on tree system integration with conventional livestock, horticultural and broadacre food production in southern Western Australia (WA).

Chapter 6 – Bioethanol is one of themost important alternative fuels nowadays. Brazil and USA are the two main producers. In Brazil, it is produced from sugar cane and used in modified internal combustion engines, in vehicles called "flexfuel cars". Biohydrogen can also be produced from sugar (sucrose) and used as a fuel in a modified internal combustion engine. The advantages for hydrogen are zero carbon emissions and the possibility to use as a bridging technology for the introduction of proton exchange membrane fuel cell (PEMFC) vehicles. An analysis based on energy content, emissions, and production yield from sucrose, for each fuel, is presented here, along with the case of Argentina, where compressed natural gas (a gaseous fuel) is commonly used in the mass transportation system.

Chapter 7 – In a process known as anaerobic digestion (AD) a consortia of microorganisms degrades organic matter to digested sludge, carbon dioxide and methane. During this process hydrogen is both produced and consumed as an important intermediary. Hence, if you wish to produce hydrogen through biological means, it becomes necessary to intervene in the anaerobic digestion process. In this short review the authors describe how is it possible to obtain hydrogen through the fermentation of carbohydrates, describing the metabolic pathways and clarifying the destiny and function of the several molecules involved. The authors hope that the information synthesized here provides clarity to the collection of information generally found in scientific publications.

In: Bioenergy Systems, Biological Sources ... ISBN: 978-1-62417-330-1
Editor: Michel C. Allard

Chapter 1

ASSESSING THE SUSTAINABILITY OF BIOENERGY SYSTEMS

Silvina M. Manrique* and Judith Franco
Non Conventional Energy Resources Investigation Institute (INENCO) of the National Council of Scientific and Technical Research (CONICET), National University of Salta, Salta, Argentina

ABSTRACT

The use of biomass -term includes materials of organic origin unfossilized- is being promoted worldwide mainly for two reasons: i) its energy contribution and ii) involvement in reducing CO_2 emissions.

Unlike other renewable energy sources, the biomass includes a variety of resources all of organic nature but with different characteristics. Bioenergy systems based on biomass are complex because they cover various stakeholders and economic sectors, different legal frameworks, resource types, implementation options and energy conversion processes, which generate systems with particular characteristics that are included in certain social and environmental contexts.

The fact that the biomass is associated with the benefits mentioned and even many other benefits (partial replacement of fossil fuels, regional and local availability, source of carbon neutral energy, low or no cost, versatility for generating electricity, heat, solid or gas or liquid fuels) not imply that their use is inherently sustainable. The condition of sustainability - which does not indicate a fixed target but a path to a more desirable stage- depends on how a group perceives the reality and what their interests are in a given historical and temporal moment. Therefore this concept must be defined in each particular context. There should also be defined criteria and indicators to monitor the trend of the bioenergy system towards higher or lower levels the sustainability in the context analyzed.

In the Lerma Valley, province of Salta, Argentina, where there are different resources of biomass, four bioenergy systems (energy systems based on biomass) were defined for analysis. By means of a Multi-Criteria Assessment, the four systems were

* Corresponding author. Tel.: +54 387 4255533; fax: +54 387 4255489. E-mail address: silmagda@unsa.edu.ar (S. Manrique)

evaluated in issues identified and agreed by means of a survey of international experts. Three of the energy systems on the basis of biomass show a level of sustainability between medium and high (50 to 75% of a perfect system). Though no bioenergy system reaches 100% of sustainability, they present valuables qualities for its implementation.

Analyses identify areas with greater shortcoming and recognition of the dimensions that require the application of policies or corrective tools. The simultaneous and planned application of the proposed bioenergy systems could have a positive impact on local sustainability of the Lerma Valley.

1. Introduction

The use of biomass for energy involves a series of links or stages, that range from the production of the resource to the processing and application of it (Giampietro et al., 2006; Buchholz et al., 2009a). These stages vary depending upon the biomass resource considered, the use it will have, the energy conversion technology that will be used, the energy distribution channels and the site where the alternative will be executed, as well as its environmental, cultural, economical and institutional characteristics in a given moment in time (Lewandoski and Faaij, 2006; Loken, 2007; Sheehan, 2009). In this sense, the contribution of the biomass as a source of renewable energy and/or carbon reserve –as two of its most appreciated features today (Hatje and Ruhl, 2000; Karekezi et al., 2004; Demirbas, 2009)– must be evaluated individually within each temporal, human and territorial context, where the management and its potential are planned (Tsoutsos et al., 2009).

The experiences on a national level, as well as the discussions and experiences recorded from different sectors of the world (Kaimowitz and Thiele, 1999; Mol, 2007; Koh, 2007; Bush, 2008; Searchinger et al., 2008; Righelato and Spracklen, 2007; Bindrabán et al., 2009; among others), lead to consideration and rethinking some basic issues related to the use of biomass resources.

Firstly, the potential of "biomass" considered as a universal category, cannot be evaluated unless the scope and the definition are specified. It is not possible to use the positive atributes of a type of biomass resource for a generic defense and especially not for its evaluation (Honty and Gudynas, 2007). It is necessary that the specific problem for each type of biofuel (fuel derived from biomass resource) whether solid, liquid or gas, gets its own identity in order to allow for the study and optimization to reach the best results in its use.

In the second place, even by analyzing the resource and specifying its production method, the benefitial potential will differ depending on the scale in which the resource will be utilized, as well as the context in which its use is planned (Mol, 2007; Koh and Ghazoul, 2008; Bindrabán et al., 2009).

Third, the condition of renewability of the biomass resources is "conditional" (Ryan and Openshaw, 1991), and if its rate of use overcomes its renewal rate, may cause deterioration of the base resource (Sasaki et al., 2010), having a major impact when it comes natural ecosystems (Costanza et al., 1997; De Groot et al., 2002; Kumar and Kumar, 2008). Fourth, given the complexity of the bioenergy systems in general, the evaluation of its "contribution" at a sustainability level in a given place, must be as inclusive and holistic as possible. What is sustainable, associated with desirable conditions in different facets of the human life, must be locally defined (George, 1999; Richardson and Verwijst, 2005; Schubert and Blasch, 2009).

Acomplishing benefits in an issue like energy provision, for example, should not generate new problems in other areas, like pollution or natural resources degradation; affect lifestyles or displace communities; subjugate human rights; lack of food or an increase in their price; weakening of the regional economies; generation of contaminating or greenhouse gases (GHG) emissions, among the main aspects (Naik et al., 2009; EUBIA, 2009; Sanhueza, 2009). The biomass, as a source of renewable energy, but also as locally appliable strategic tool, could contribute to the establishment of more sustainable energy systems, which should be evaluated in each case. Particularly in the northwest of Argentina, which includes the Province of Salta, where it has been identified as a territory or region with a wide range of biomass resources (GENREN, 2007).

For the generation of successful proposals, projects and activities that include bioenergy systems, it is necessary to understand that reality –whose systemic nature is complex in itself – does not lead "downwards", in lower levels of the system, at a basic simple level. Even the simplest or elemental parts of reality are conformed by interrelations among its components, forming part of organized groups (Morín, 2006). The reality, that manifests in multiple levels and with the same ontological category, must therefore be understood and explained simultaneously, from all possible perspectives.

2. The Sustainability As Analytical Framework

2.1. Concept, Origin and Importance

As a concept, the sustainability has a long and rough history. Its roots go back to the controversial writings of Thomas Malthus (dated 1798), who dares to suggest that the planet has reached the limits of its capacity to maintain or support the human population and the needs of the society (Sheehan, 2009).

The United Nations have made the term "sustainable development" popular when they wrote "*the sustainable development is that which satisfies the needs of the present generations without compromising the future generations*" in the Our Common Future Report (1987:43), also known as the Brundtland Report (WCED, 1987). The WCED report is recognized as the first global and institutionalized attempt to simultaneously work with the economical welfare, the environmental protection and the social justice, mentioned as the three pillars of the sustainable development (Tijmes and Luijf, 1995; Lumley and Armstrong, 2004). This report reaches the state of reference point for each debate and search for a redefinition of the concept (May and Brennan, 2006; Holden and Linnerud, 2007).

While the sustainable development, at least in theory, is a common target for the nations, communities and businesses, the definitions still appear diffuse and with yet no scientific agreement (Phillis and Andriantiatsaholiniana, 2001; May and Brenan, 2006; Sheehan, 2009). Furthermore, there is a wide variety of definitions and interpretations, for which some scientists avoid using the term due to its vagueness (Mebratu, 1998). Luke (2005) suggests that the concept is increasingly being used as a fashionable label while it is still directly associated with a development model where the economic growth is prioritized (Holden and Linnerud, 2007). Certainly, the macro-economic problems, the techno-industrial civilization ones, the manner and life gender homogenization problems, the destruction of the ancient farming world, the pollution, the generalized urbanization, the destruction of solidarity, the drugs, the alcohol, show the imperious need to redefine the concept itself of development (Morín and Kern, 2002).

The sustainability, apparently mentioned as a synonym of sustainable development, is, or could be, a totally new concept (Seghezzo, 2008; Sheehan, 2009) and some fundamental distinctions have been found in both terms. The main distinctions are seen, without a doubt, in the ethical rethinking, and the search to disassociate the concept of development with its strong economical connotation (Times and Luiijf, 1995) and its excluding and dichotomist posture (Bell and Morse, 2005). In this context, the sustainable would include aspects like: an acceptable degree of political and economical decentralization, social justice, strengthening or empowering the local communities, recognition and valuing the importance of the cultural differences, and each nation's right to execute an effective domain of ecosystem goods and services (George, 1999). This means the well being of the mostly of people in the greatest number of areas that are a part of them: moral, intellectual, cultural, social and physical. This also includes the future generations.

There is no universal sustainable condition toward which everything should be directed. What is sustainable has a clear particular connotation in each case, context, moment and with every human group, each in which its essence and "utopia" to follow is defined (place, permanence and persons, according to Seghezzo, 2009). If the sustainability is correctly defined in each of these situations, and verified and permanently reoriented in its course depending upon the observed tendencies (George, 1999; Ness et al., 2007; Mendoza and Prabhu, 2003) surely, the unsustainable will slowly begin to be left behind.

2.2. The Energy Sustainability: An Integrating Perspective

In the study of biomass as a source of renewable energy but also as a physical matrix that allow the emergence of ecosystem functions, it was considered necessary to coin a particular connotation for the sustainability. Considering the intercrossing of variables, factors, dimensions, actors, interests, possibilities, resources, that were related to biomass, it was necessary to define a practical and useful analytical framework that allowed the detection of the fundamental aspects of the selected territory in which the proposed bioenergy systems could impact; the meaning of "sustainability" from the involved human groups's point of view; and the direction towards which the studied "system" should be oriented (Lerma Valley, province of Salta, Argentina).

From the paradigm of sustainability, there are a series of elemental questions: What is the relationship between bioenergy and sustainability? How can be applied the concept of sustainability in the development and implementation of the bioenergy systems? How can it be evaluated if the bioenergy systems proposed for the Lerma Valley constitute valid options towards the implementation of more sustainable energy systems? How much can biomass contribute to the transition into integrally sustainable energy systems?

The link between the sustainability aspects and the biomass as a study object were reflected in the concept "energy sustainability". It started from the recognition of the importance of the energy within any type of system; its fundamental role in basically all aspects of human life, including access to water, agricultural productivity, healthcare, education, creation of jobs, resources and the environment (Takada et al., 2000; Kalitventzeff et al., 2001; May and Brennan, 2006); and therefore, its potential to constitute in an analytical framework through which all the elements and components of the system, territory or unit of analysis could be evaluated. Elements and components that were connected with different energy aspects.

The idea of an energy system as an analytical framework is to see the world as energy fluxes where each human activity is linked by these fluxes and at the same time, with the environment. People use energy derived from natural resources which when burned release heat to the environment (Kern and Smith, 2008). The energy is not consumed; it is transformed into different types of energy at different levels of the energy supply chain, starting with the solar energy used by the plants (Kemmler and Spreng, 2007). Furthermore, the use of energy produces environmental impacts (Hammond, 2004). This way, the process of human development is closely related to the evolution of exploration and domain of the use of the energy sources present in nature and the structuring of complex energy systems (Paz et al., 2007; Haas et al., 2008).

The idea of using this framework to relate a group of sustainable development indicators is relatively new (Kemmler and Spreng, 2007). This analytical framework presents a series of specific advantages (and some restrictions in terms that some aspects probably can't be evaluated) like for example, that most subjects related to the environmental sustainability (like one of the facets of sustainability) are related to the use of energy. For this reason, the indicators selected from the base of the energy can cover many of the relevant environmental sustainability issues. Energy is also linked to social aspects such as the access to energy sources, energy dependency, possibilities of social and personal realization, use of time or healthcare. Lastly, the demand for energy in the developing regions is an essential problem for the economical development (Hammond, 2004; Demirbas and Demirbas, 2007). Dincer and Rosen (1998), note that the energy resources are closely linked to the sustainable development. It is possible, for example, to compare different development models by analyzing the benefits of a larger energy coverage or fuel for the productive or industrial development, versus the benefits of preserving the forests or planting energy crops, in terms of GHG emissions. The local participation can guide the decision making, also incorporating into the analysis the different social values like the community integrity, identity, freedom, happiness, security, gender equality, auto-determination (Elghali et al., 2007).

The construction of the concept of sustainability from an energy point of view requires taking into account aspects such as: the prevention and reversal of environmental impacts that result from the current energy production and consumption system; the security in the supply of the different fuel sources and the decrease in energy dependency (especially the sources that generate high social and environmental costs); the coverage and the entire population's equitable access to the energy resources and services; the population's democratic participation in the decision making processes about politics and energy projects, as well as the different available technological options (Berman et al., 2004; Bertinat and Salerno, 2006; Evans et al., 2009). The energy sustainability demands consideration and satisfaction of the society's needs, not only of physical survival but also economics, social, political, cultural and environmental needs for a dignified life (Takada et al., 2000; Haberl et al., 2004; Buchholz et al., 2009 a and b). The type of energy used, the way it is produced, the access and distribution mechanisms and the benefits determine the type of development and the levels of sustainability possible to achieve (Takada et al., 2000).

When talking about more sustainable energy systems, the reference is not exclusive to the type of sources and processes involved in the generation of useful energy, or just the technological aspects. All these aspects are considered in conjunction to the impacts and connections with the territorial, temporal and human dimensions. This implies the inclusion and consideration of the physical and cultural contexts (including every aspect that makes life human), with its opportunities and strengths, weaknesses and threats. The biomass is part of the solution to fight against global warming, as well as to provide creative solutions to local level. But it is not “the” solution, and it could even become part of the problem that it is trying to resolve, or even generate new problems. The energy sustainability of the territory, properly defined, achieved with consensus and sustained within ethical, fraternal and supportive values will allow for unsustainable energy systems to be left behind.

2.3. The Context of Bioenergy in Argentina: Main Controversies

The biomass (only considering wood and bagasse) has been, in relative terms, losing its participation in the national energy matrix, fluctuating from over 4% in 1970 to approximately 2% in 2005 (in a total of almost 72,000 toe/year). There is also an addition of an uncertain participation of other biomass resources that are categorized under “other primaries”, like agrarian residues, quebracho tree saw-dust, black liquor and others (which represent 0.8%). However, the National Secretary of Energy informs that, based on the data provided by the Forest Department, there is a lack of 70% of the information for the year 2006 corresponding to firewood as such, derived from the lack of information from the provinces of Chaco, Jujuy, La Pampa, Salta, San Luis and Santiago del Estero (SEN, 2006). Therefore, the final numbers are fairly higher to the ones mentioned. In the same way, the WISDOM project (FAO, 2009) mentions that, because the National Energy Balance (NEB) concentrates on fuels used for the auto-production of electricity (which represent about 2,255 toe/year, and which include the category “other primaries”), the total real consumption of biomass for energy within the country is close to 5,000 toe/year (about double than the number considered by NEB).

Beyond the real current contribution of bioenergy, given the existence of different natural ecosystems in the country (PNUMA, 2004) as well as the agricultural-livestock characteristics of its mains economical practices (SAyDS, 2005), there is a large potential for the use of biomass resources. This potential must be considered in two separate facets. On one hand, the woody biomass (firewood) generated in the natural ecosystems, has been traditionally used by the population without any type of forest management. The lack of plans for the resource management, the intense wood exploitation and the national administration being managed by a sector with antagonic interests (the agricultural and livestock farming sectors) have gradually deepend the deterioration of the forests and the natural ecosystems (SAyDS, 2005). On the other hand, and precisely because of the agricultural characteristics of this country, one of the main biomass resources being currently promoted -and whose surface has been expanding in the last years- is the soybean farming (Panichelli et al., 2009). Argentina is one of the three greatest vegetable oil producers and exporters and the largest global soybean and sunflower oil exporter (FAO, 2008). Given that the production of biodiesel (diesel obtained from vegetable oil) is just one more link in the oilseed chian (CAER, 2008), in 2007, the country positioned third on the list of countries with potential to produce export quality biodiesel (Johnston and Holloway, 2007). 90% of the existing biodiesel plants are large scale (with a capacity of over 33,000 m^3, CESPA, 2007). The advantages of the scale economy (which also allows the possibility to negotiate the price marking), leaves the domestic and small plants out of the competition (Tomei and Upham, 2009). In 2008, Argentina produced over 10% of the world's biodiesel (Lamers, 2006). Reaching this position meant that in the years 2000-2008 the total cultivated area increased 32% going from 24.5 to 32 million hectares (FAO, 2008), where the soybean occupied 50% of the grain covered surface (Panichelli et al., 2009). In consequence, the loss of native woodland associated with the advance in the farming frontier (as one of the main impact factors, among which also include livestock activities) has reached critical levels in the last years (PNUMA, 2004). The production of biodiesel has created controversy not only at a national level, but also internationally mainly due to the competition in the use of the land (Righelato and Spracklen, 2007; Searchinger et al., 2008; Bindrabán et al., 2009; Phalan, 2009); competition for the water (Schnoor et al., 2008; Domínguez Faus et al., 2009); its possible repercussions in the prices of food products (Rahman et al., 2008; Koh and Ghazoul, 2008); its effects in the biodiversity (Fitzherbert et al., 2007; Koh and Wilcove, 2007); its energy balance (Pimentel et al., 2007; Rajagopal et al., 2007; Campbell et al., 2009) and the not always favorable balance of GHG (Farrell et al., 2006, Hill et al., 2006, Alder et al., 2007; Gibbs et al., 2008).

The main limitations for the bioenergy, unlike what was detected for renewables in general in Argentina (Fundación Bariloche, 2009), aren't political-institutional or economical-financial barriers. Although these aspects are fundamental, there are some identificable important points that must be evaluated in any project in this issue. Among them, some can be mentioned: i) aspects related to the biomass resources management (supply levels required for the production, existence and types of management plans, feedstocks manipulation, type and intensity of the required machinery, for example); ii) the scale of the projects (extensive monocultives or prodial crops, for example) and iii) the use and destination of the generated energy (export or local destination and consideration for the priority energy demands). These are probably the three aspects that currently generate controversy with the promotion of the bioenergy in the country.

As has been mentioned before, there are on a national level different biomass resources as well as uneven geographical distribution of them. The potential for each of these resources should be considered within the three highlighted aspects. It is not possible to talk about "biomass" as a whole when each type of resource demands an individual analysis, is located in a particular place, is generated under determined conditions and could have a differential reach. The highest expectations in the biomass field currently fall onto the food energy crops for the liquid biofuel production, which could imply, among others, diversity and security in the energy supply (Lamers, 2006; USDA, 2008; Lamers et al., 2008). The controversy arise mainly when discussing about the impacts the expansion of the dedicated energy monocultives (Sarandón and Iermanó, 2005 a and b; Salomon et al., 2006; Seijo, 2008; Van Dam et al., 2009; Panichelli et al., 2009); the high supply and resource dependency that it implies (water, land, fertilizers, machinery) (Seminó et al., 2009; SAyDS, 2008; Tuesca et al., 2007; Rodríguez and Jacobo, 2010), the biofuel's destination (Lamers, 2006; Tomei and Upham, 2009), and the type of impacts on the region where it is produced. In Argentina especially, regarding biodiesel of soybean.

The use of a technological package, consistent of genetically modified seeds (GM), zero tillage and glyphosate, have consolidated an agricultural model focused on exports, based on the mechanized production and large scale. This technology package has allowed the expansion of crop to areas that were previously inadequate, generating great pressure on the marginal lands and non agricultural areas (Qaim and Traxler, 2005; Zak et al., 2008). Today, the GM soy accounts for over 98% of the soybean produced in Argentina (Tomei and Upham, 2009). However, the intensive use of agrochemicals (glyphosate), which has increased almost 161 million liters in 10 years (SAyDS, 2008), has implied consecuences on the physical (Duke and Powles, 2008; Powles, 2008; Binimelis et al., 2009; Rodríguez and Jacobo, 2010) and social environment (Salomon et al., 2006; Seijo, 2008; Seminó, 2008). The scale economy inherent to the new production model has driven the concentration of wealth and property of the land (Giarracca and Teubal, 2002; Joensen et al., 2005; Robin, 2005; Lamers et al., 2008). The intensification of the agriculture has guided the reduction of the rural labor force, since small farms can generate one job for every 8 hectares, the mechanized plantations employ one person every 200 hectares (Carvahlo, 1999; Dros, 2004). These processes have reinforced the rural uprooting and exodus (Tomei and Upham, 2009). If the international price and biodiesel consumption situation is favorable in the long term, one can estimate that there will be a demand in expanding the agricultural frontier in order to obtain more land for the soybean cultivation (Bertinat, 2007).

A recent study by Panichelli et al. (2009) about the Argentine production of soybean biodiesel for export shows that from the GHG emission's point of view, the biodiesel production is not a good option for mitigating global warming, mainly due to the land-use change (forests to agricultural lands and subsequent changes in the carbon stocks). Van Dam et al. (2009) state that the soybean production for biodiesel would be sustainable if it were produced in abandoned agricultural territories. Tomei and Upham (2009) mention that the exponencial increase of soybean and its exportation, which prioritize the short term benefits, has produced a system that is not economically and environmentally sustainable in the long term. Sarandón and Iermanó (2005a) reveal that the surface that will be sowed with the main oleaginous crops, according to the chosen crop and the associated technology, to fulfill the set targets in the national politic (Law 26093), is largely superior to the currently cultivated. The afore mentioned authors state that the expectations for the use of the biodiesel as a replacement for the fossil fuels do not appear compatible with the ecological constraints that are long-term.

This outlook enables the identification of the highlighted aspects: resource management, destination and scale-application, as the main issues that must be discussed and agreed upon before promoting new initiatives in this field. The reigning controversy, both globally and nationally, oriented the options done in this investigation where the dedicated energy crops are only considered from the natural existing formations, and oriented towards degradated places. Until there are more investigations about first generation liquid biofuels (based on sugar or oleaginous crops) that are currently being developed, they do not constitute a feasible option for elevating the quality of life of the Lerma Valley, nor contributing to the implementation of more sustainable energy systems. However, other biomass resources, whose contribution is restricted to the place where they are found, could constitute in valuable elements in the implementation of local development strategies. Some of these alternatives are explored in the current chapter, analyzing in what way and in what aspects each of these potential bioenergy systems could impact the level of local sustainability.

First it explains with what criteria the bioenergy systems were selected for the Lerma Valley. Secondly the local context for the implementation of these bioenergy systems is described. Finally, the systems are compared in sustainability criteria defined by an expert survey.

3. Sustainability of the Potential Bioenergy Systems for the Lerma Valley, Salta, Argentina

3.1. Available Biomass and Energy Supply

The Lerma Valley, in the center of the Salta province, north of Argentina, is an intermountainous tectonic depression which is located near the last foothills of the Oriental Mountain Range to the west and the Sub Andean hills to the east. The average altitude is 1,000 meters above sea level. The total surface, considering the water division line, is approximately 500,500 ha with a maximum length of 144 km and a maximum width of 53 km. It is located among the coordinates 24°22.0' a 25°43.0' latitude south and 65°15' a 65° 48' longitude east (Núñez et al., 2007).

The Valley is composed of 7 departments and 13 municipalities. Of the 1,084,033 inhabitants in the province in the year 2011 (INDEC, 2011), 53% is concentrated in the Lerma Valley, although 44% is in the capital department. Excluding the capital, 70% of the Valley population corresponds to urban population (with over 2,000 inhabitants) and 30% to rural population (this includes the rural population grouped in towns of less than 2,000 people and to the rural population dispersed or residing in open countryside).

Considering the Potential Biomass (PB) or total biomass in the Valley (Table 1), it was necessary to define criteria (Manrique et al., 2011) that allowed to discriminate most promising resources for energy uses (Available Biomass or AB). It was assumed that the local energy demand had priorities from the standpoint of sustainable development. It was necessary to reduce the possible universe, concentrate efforts, and optimize scarce resources (time, money, personal), and define specific bioenergy systems that could be developed in the Valley. From the analysis of the available international literature and from personal considerations, eight criteria were formulated (Table 2). Criterion 1, "Existence", allows from the base information to have a first approach to the quantification of the resource, which will be needed in later studies, for each of the resources identified as AB.

Criterion 2, "Equilibrium", incorporates into the analysis the frequency with which the resource is generated. If the generation frequency of a resource is punctual, i.e. once a year, or once a month, in a specific place, it is impossible to think of facilities that depend on this supply, and forces the complementation with other nearby resources. It is necessary to evaluate this parameter before moving on to greater studies. The observation of the generation frequency also marks the natural sustainability of the resource in time, which will also depend upon the level of anthropoid intervention that enables or not the continuity of sustainability.

The scattering (criterion 3), or on the contrary, the concentration of a resource in a specific place, makes it potentially useful for one or another sector since its energy benefits will not reach the majority of the Valley population, which is the target in this investigation. Now, a type of biomass that is too scattered could result in high costs for its use, which could also be evaluated. On the other hand, depending upon the scale with which it is analyzed, a resource that is concentrated in a geographic region could be beneficial. In this case, the evaluation took into account the amount of municipalities the resource was concentrated in within a 30 km radius, considering that a high local scattering is unfavorable (less score), since it implies higher costs for transporting that resource to the place where it is used.

Table 1. Potential biomass identified in the mountain area and low area of the Lerma Valley. Where: X: presence; X_C: commercial; X_S: subsistence

Type of biomass	Area of the Valley	
	Low area	Mountain area
Natural biomass		
Phytogeographical province Yungas	X (Foothill)	X
Phytogeographical province Chaco	X	X
Phytogeographical province Prepuneña		X
Other forest lands: shrubland	X	
Agricultural residue biomass		
Industrial crops	X	
Legume	X	
Perennial forage	X	
Annual forage	X	
Grain cereals	X	Xs
Fruit trees	X	
Vegetables	Xc	Xs
Other crops	X	
Residual forest biomass (only from plantations)	**X**	
Forest		
Urban residual waters (sewages effluences)	**X**	
Industrial effluences	**X**	
Food		
Beverage		
Other		
Stabled livestock waste		
Cattle	X	
Bovine	X	
Caprine	X	
Equine	X	
Porcine	X	
Ass/Mule	X	
Free range livestock waste		
Cattle	X	X
Bovine	X	X
Caprine	X	X
Equine	X	
Porcine	X	X
Ass/Mule		
Urban Solid Wastes (USW)	**X**	**X**
Mud sediments (dams, reservoirs)	**X**	
Mud from sewage treatment plants (EDAR)	**X**	

Table 2. Criteria to select the Aavailable resources of Biomass (AB)

#	Criteria	Definition
1	Existence	Amount in weight or volume, of the resource in surface unit.
2	Equilibrium	Periodicity, frequency or resource generation rate by day, month, year or production cycle.
3	Dissemination	Resource concentration. Punctual source of generation.
4	Effects	Visible local impact. Its use can visibly impact the region in terms of contamination, emissions, residue disposal, or others, for which its use is proposed or discarded beforehand. Social or other factors can be taken into consideration.
5	Experiences	Current uses. Established uses of the resource like for example, fertilizer, forage, crafts, etc.
6	Requirements	Legal and physical accessibility. Includes individual ownership, protected areas legal restrictions, and other. Orography is also considered – difficult terrain situations.
7	Estimates	Perception of the resource and the interest of its use or social acceptance (adaptable).
8	Expectations	Perspective of the future existence of the resource in quality and quantity. Probability that the source of generation of the resource will continue to exist, either by political, legal, economical, financial, cultural, or other backup.

The impact of the use of that resource (criterion 4) includes a preliminary evaluation about the possible environmental, social and economic impacts. It can be very easily perceived that the larger the expectations for positive impact, the higher the score. These impacts can be recognized at this stage with office studies, terrain recognition and/or local interviews. However, those which appear to be the greatest advantages in the use of a resource might turn into the worst impacts of the region. It all depends upon how the proposed systems are defined, planned and executed.

Different traditional uses of the resource (criterion 5) can overlap resulting in an over-demand of the resource and therefore, resulting in a greater pressure on them. It is necessary to identify the uses and observe the adaptation of the project in the area. In this case, the higher the demand for the resource (for other uses), the higher score. The current use of the available resources can absolutely limit the use of a certain resource, like for example, if a harvest residue is the feedstock for a handcraft factory, or if they are the base for the next sowing cycle.

Regarding the demands (criterion 6), defined as restrictions of the human or natural systems, they can confine the available resources even further since the owner may not share the idea of its use, or the orography (by which the study area has been divided into two sectors: low and mountainous) may not allow the transportation of the material to certain places. The restrictions must be carefully analyzed.

At greater restrictions lower score. However just one restriction for example legal, could be enough to avoid the use of the resource. Therefore, not only how many but also which restrictions exist that should be identified. The perception of the local people on the resource and its use (criterion 7) is of fundamental interest to know the possibility of the permanence of the project in the time and its social impact. If there is local acceptance, it will be a long lasting project and for this it gets a higher score. Generating any shift in the cultural standards is a costly process in terms of time and effort, but trying to impose something that is not culturally accepted or recognized, will undoubtedly imply an unproductive loss of time.

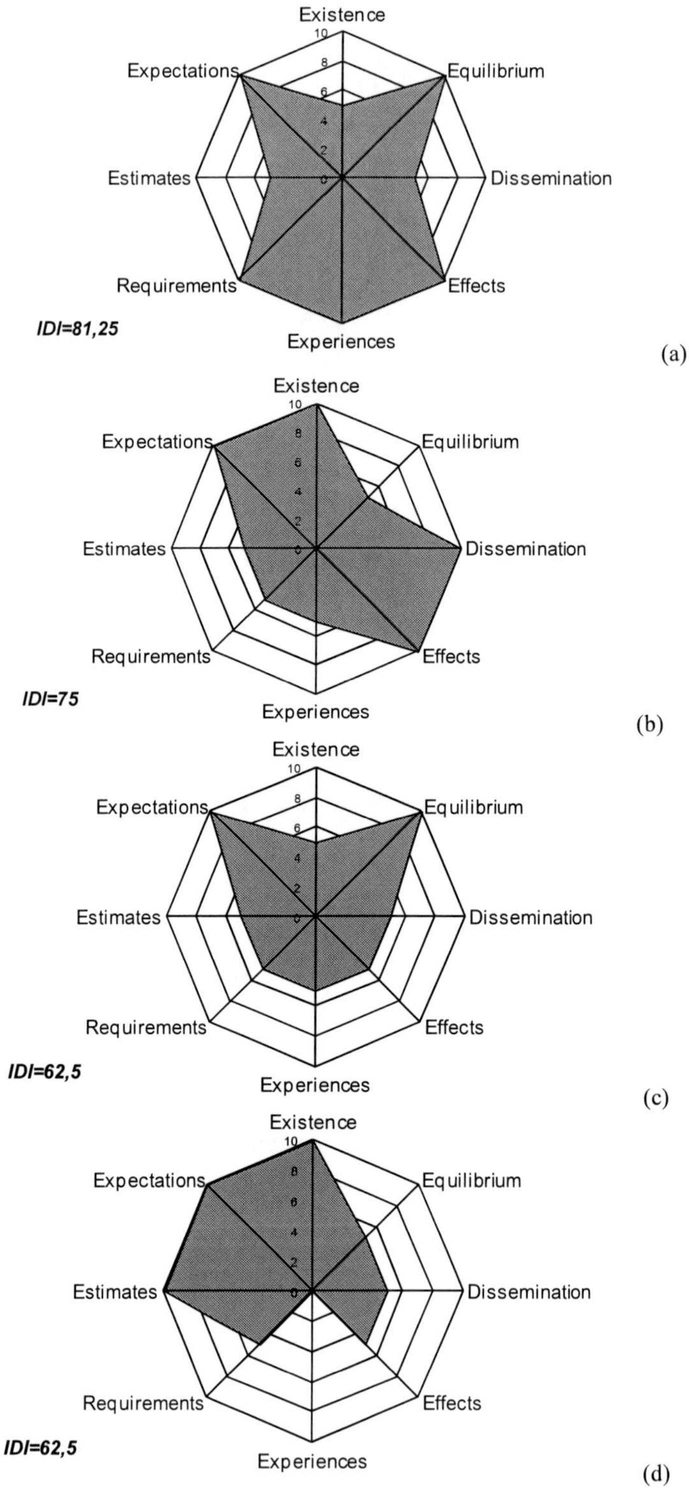

Figure 1. Representation of the Availability Index (AI) of the main biomass resources in Lerma Valley: (a) Urban Solid Waste (USW): *optimal resource*; (b) Virginia tobacco residues: *stable resource*; (c) Criollo tobacco residues: *stable resource*; (d) Acacia spp timber biomass: *stable resource*.

Finally, in terms of expectations (criterion 8) that the resource will exist in the future or the probability that the source of generation of the resource will continue existing, it is notorious that if the probability is high, the score will be higher. If this source will disappear the following year, only short term exploitation can be applied, even with the restrictions that there might not be enough time for the resource to be studied and its use planned. This criterion is of fundamental importance for the projection of its use, especially on a large scale, due to the level of investments required. However, it is also important on a smaller scale because if awareness, training and learning are projected, and the source of generation disappears and therefore the resource itself, it would imply a waste of valuable time. Furthermore, the frustration among the participants, and the future unwillingness to participate in any similar proposal should be added. A qualitative categorization was defined for each of these criteria with a low, medium and high scale (Table 3) and a number value of 0, 5 and 10 respectively, where 80 is the highest score that every resource can obtain.

The "Availability Index" (AI) incorporated the values of each criterion according to analyzed resource, and was multiplied by 100 (Manrique et al., 2011). Every PB resource was classified according to its AI, and the ones which obtained a value of over 50% were chosen.

The biomass resources that show a high AI (Figure 1) can be grouped into two categories (Manrique et al., 2010a): a) residual biomass: agricultural residues from three crops (Virginia and Criollo tobacco and pepper) and urban solid waste –USW– and b) natural biomass: woody species from the shrubland (degraded aspect of the Chaco ecosystem defined by Cabrera, 1994).

Other important ecosystems of the area are the Yungas and Chaco (Cabrera, 1994). The bioenergy systems that have been defined as the ones with the highest possibilities of application according to the demands, the existing resources and their potential and the available technologies are what is described below (see more details in Manrique et al., 2008; Manrique et al., 2009; Manrique et al., 2010 a and b; Manrique and Franco a, in press):

1. *Urban Solid Waste (USW):* Approximately 24,100 t (humid weight) is generated annually in the Lerma Valley, with a potential energy of 9,000 Gcal/year. The potential has been estimated considering that the residues could be deposited in a regional sanitary landfill where the biogas, product of the residue decomposition, could be exploited (for a 70% of biogas valorization). This potential could be used for producing electrical energy from the USW produced in the lower region of the Valley, and the location for the sanitary landfill could be planned according to the electrical network and possible derivation of bioenergy from the landfill. This energy could be used for public lighting, for example. The available systems are the internal combustion motors used for burning biogas and generating electricity.
2. *Woody biomass from Acacia spp. (Acacia caven, A. aroma, A. furcatispina and A. praecox):* In the shrubland there were some more densely populated species that were found to be potentially adequate for their management as opportunity crops (i.e., to make use of their annual energy supply in the same place where they grow). This bioenergy system is based on the biomass supply of the four Acacia species that were studied in the Valley (Manrique et al., 2008, Manrique and Franco b, in press). These species are scattered in about of 68,000 ha and in many cases, they have formed almost pure forests. The potential energy is 50,900 Gcal/year from an annual wood supply of 15,500 t (dry weight). It is proposed to use the biomass for its

combustion in efficient furnaces and ovens, mainly for the tobacco sector of the Valley what is fuel-consuming (demands large amount of fuel for the tobacco drying furnaces and other instances of the productive cycle).

3. *Agricultural residues (Virginia and Criollo tobacco, and pepper residues):* There is about 46,000 Gcal/year available from 18,000 t (dry weight) of residues generated from these crops. The annually cultivated surface is of 17,300 ha on average. It is proposed the use of these residues for the tobacco drying furnaces (or other caloric demands from the productive cycles of these crops) replacing natural gas or firewood obtained from the region's ecosystems which don't have a planned management. The technologies available for the use of these residues are fixed grill boilers.

3.2. The Local Context for the Implementation of Bioenergy Systems

The measurement of sustainability is the strongest conducting force for the discussion of sustainable development. A reliable measuring tool for sustainability is a prerequisite for the identification of the processes, alternatives or non-sustainable systems (Afgan and Carvalho, 2008). For the Lerma Valley, before analyzing the potential bioenergy systems that could be implemented in the area, it was necessary to know the baseline, in order to be able to monitor changes of the system as a whole.

It was the WCED (World Commission on Environment and Development, 1987) who introduces the term "sustainable development" (SD) in the international agenda. It is recognized as the first global attempt and institutionalization of treating the economic welfare, the environmental protection and the social justice simultaneously, taken as the three dimensions of development (George, 1999). This report suggests a group of standards and becomes a point of reference for every debate on sustainable development (Holden and Linnerud, 2007). The SD depends on society's points of view and values (Sheehan, 2009), since these define which matters are considered important, which issues are investigated and what objectives must be pursued. The future generations, with greater knowledge, more sophisticated technology and different needs, will define SD in their own terms and with other objectives. However, despite the differences in opinion on the subject (Mitcham, 1995; Mebratu, 1998; George, 1999; Mauerhofer, 2008), there is an agreement that the SD is not a fixed condition or a final state, but it is a dynamic process (Mog, 2004). Although nobody can know how a sustainable society will be in the future, it is clear that some human activities, if not stopped or changed, will not contribute to the achievements of that society (Seghezzo, 2008).

Kemmler and Spreng (2007) mention that this dynamic process must be continuously evaluated: the system's tendencies and the desired goals must be evaluated and corrected accordingly. George (1999) states that its measurement is indispensable for the operationalization of this concept: it should be possible to measure whether a type of development is sustainable. This requirement was recognized in the Agenda 21 which states that the sustainable development indicators need to be developed in order to provide solid bases for the decision making at all levels. The indicators facilitate the orientation in a complex world by condensing great amounts of information in recognizable patterns (Bossel, 1999; Afgan and Carvalho, 2008). For this, it is necessary to identify and consider each significant aspect of the environment and each important component that contributes to the quality of life.

On the other hand, the establishing indicators show a greater sensibility if they refer individually to each component that has been identified as important, since the added indicators can hide the weight of some of the integrating factors and show adverse results (George, 1999). In addition, it is necessary to know the impact each of these components has on the quality of human life and the environment in the case they are modified, since in some cases, a small modification may not be acceptable.

The multidimensional treatment of the SD is just a reflection of the complex reality of the "system". It is recognized that each dimension has its own characteristics and at the same time it is conditioned by the other dimensions (Alburquerque, 2004). The perception of the environmental, social and economic risks varies from one individual to another. The NGO, local governments and the scientific sector may prioritize the social, technical and environmental aspects, while the private producers might prioritize the economic aspects. A balance in opinions is therefore necessary to accomplish a consensual definition of the conditions that might ensure a sustainable use of bioenergy in a particular region (Gnansounou et al., 2007).

A researched done on a sample of the Lerma Valley population (Manrique et al., 2010 c) which considered the afore mentioned dimensions, shows that usually, the population recognizes the economic dimension as the one with the most weight (36%), followed by the social and environmental, each with a similar weight (25% in each case) and lastly, the institutional dimension (14%). In the opinion of the consulted persons, the economical aspect is a priority in the sustainability of an initiative in the region because it is the local people's main interest. This prioritization has been recognized on a global level as typical among the developing regions and third world countries (Gnansounou et al., 2007). In an analysis broken down into different opinions, one can see that the companies define their projects granting the highest importance to the economical and environmental aspects, which makes sense in the current legal environmental frame and in the search of maximizing the profits. The governments spread their interest scale between environmental, social and institutional aspects. The town people and local producers give most importance to the economical, technological and social areas, pointing out the impediment the technology imposes in many experiences and the need to ensure their job and income resource. The associations define their decisive axis as the economic and technological facets, similar to the producer, which makes sense because in most cases, the producers make the associations.

The indicators that allow the characterization of the situation in the Valley (Table 4) are defined and measured in the paper mentioned before (Manrique et al., 2010 c). These indicators show, through a Sustainability Development Index (SDI) that is represented on the axis, a "baseline image" analyzed system in a multidimensional framework of SD. Each axis in the figure represents an indicator, adjusted in a way that the wider and more homogenous a shaded area, the better the situation of a system. In this case, one can infer that the system shows errors. The utilized analysis scale ranges from 0-0.2: represents a state of the system with a high chance of *collapse*; between 0.2-0.4: indicates a *critical* situation; 0.4-0.6: corresponds to an *unstable* system; 0.6-0.8: symbolizes a *stable* system, and 0.8-1 is considered the *optimal* situation of the system. According to this scale, and for an estimated SDI of 0.218 (see the mentioned study for more details in the methodology), the state of the Lerma Valley as a system can be defined as critical (Figure 2). There are great unbalances between each dimension and within them, and especially, the environmental facet is highly postponed. The analysis of each dimension allow for a highlight of the most critical variables in each of them.

As it is clearly observed, any used indicator related to the environment shows very low values, practically showing the lack of rational management of the resources and the protection and value of the environment. The level of environmental consciousness is low, although there are some possible recognizable damages due to human activities, there is no will nor decisions to seek possible solutions. Socially, one can observe large structural deficiencies of income distribution, education and health, which must be taken into account in the first place. Although there is a great enthusiasm in the participation of bioenergy projects, these projects must be approached for resolving the basic demands. Regarding the institutional-organizational, there are some strengths in the system that must be observed, the producers getting together for the irrigation associations, tobacco societies and cooperatives, local producers associations, all of which gather the community to realize common efforts towards the objects of their interest. In this case, it could be very important for the viability of the implementation of the bioenergy systems although the trust and communication with the leadership must be reinforced.

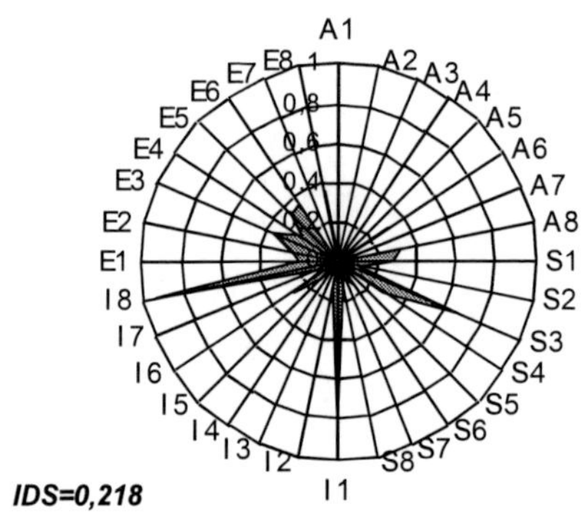

Figure 2. Current situation of the Lerma Valley according to the SDI (Sustainable Development Index): "*Critical situation*".

Economically and productively, there is a strong dependency on subsidies and financing mechanisms which could point to a weak economy. The local capitalization levels are very low and there are large discrepancies with the income, deficiencies in the productive infrastructures and a high social unconformity. This analysis allows for a visualization of the problems and system unbalances in just one image, and therefore point towards the dimensions and variables which require the implementation of corrective policies and tools.

The initial diagnostic of the Valley situation regarding the sustainability level as a system may be improved through the implementation of the mentioned bioenergy alternatives, which would elevate the previously estimated indexes. In a general view, the bioenergy projects would imply a better situation or a more sustainable situation than the current one. Great part of the considered indicators could be more or less impacted (however, structural deficiency aspects require more complex and integral political decisions). The greatest change could be noticed in the environmental dimension due to the preliminary estimates of the benefits the

Table 3. Parameters that define the categories in the scale, according to each criterion for the analysis of the Lerma Valley as a system. The letter "C" in the first column means "criterion"

C	High	Medium	Low
1	The resource is generated or exists in over 8 towns of the Valley.	The resource is generated or exists in 4 to 8 towns of the Valley.	The resource is generated or exists in less than 4 towns of the Valley.
2	The frequency of generation of the resource is daily to monthly.	The frequency of generation of the resource is monthly to annually.	The frequency of generation of the resource is biannual or multi-annual.
3	The resource is concentrated (in a 30km maximum radius from the center of ach town) in over 8 towns of the Valley.	The resource is concentrated (in a 30km maximum radius from the center of ach town) in 4 to 8 towns of the Valley.	The resource is concentrated (in a 30km maximum radius from the center of ach town) in less than 4 towns of the Valley.
4	The impact of the use of this resource would reach the population of 8 or more towns in the Valley (although not all its inhabitants).	The impact of the use of this resource would reach the population of 4 to 8 towns in the Valley (although not all its inhabitants).	The impact of the use of this resource would reach the population of 4 or less towns in the Valley (although not all its inhabitants).
5	There are no other uses for the resource in the Valley	There are 1-2 identified other uses for the resource in the Valley.	There are 3 or more identified other uses for the resource in the Valley.
6	There are no detectable legal or physical restrictions for its use in the Valley.	There are 1-2 legal or physical restrictions for its use in the Valley.	There are 3 or more legal or physical restrictions for its use in the Valley.
7	There is acceptance and/or interest for the use of the resource in inhabitants of over 8 towns in the Valley.	There is acceptance and/or interest for the use of the resource in inhabitants of 4 to 8 towns in the Valley.	There is acceptance and/or interest for the use of the resource in inhabitants of less than 4 towns in the Valley.
8	There is a high probability that the resource will continue generating (due to political, legal, financial, cultural and other factors) in at least 8 towns of the Valley.	There is a probability that the resource will continue generating in 4 to 8 towns in the Valley.	There is a probability that the resource will continue generating in less than 4 towns in the Valley.

Table 4. Criteria and indicators used for the analysis of the SD in the Lerma Valley (Manrique et al., 2010 c). The letter "I" in the first column means "Indicator", while that the following letters mean: A, environmental; S, social; I, institutional-organizational; E, economical-productive

I	Criteria	Indicators
A1	Residue management	Percentage of daily waste (per capita) recycled or treated with some sanitary system.
A2	Water management	Percentage of the population which manifests provision, treatment and/or distribution problems for the development of their activities.
A3	Soil protection	Quantity of producers with soil and fertility loss problems.
A4	Protection of the biodiversity	Number of producers with diversified production.
A5	Use of energy renewable	Percentage of the use of renewable energy sources over the total consumption of energy in the productive cycle.
A6	Natural resources management	Percentage of the population that uses fuelwood (direct or indirectly without a management plan).
A7	Local contamination	Number of producers who use fertilizers or pesticides.
A8	Value of the natural environment	Quantity of people who are aware of problems related to the environment.
S1	Population with low income	Percentage of producers who depend on the family labor (including children under 16).
S2	Home healthfulness	Population with access to sewage services.
S3	Accessibility to modern energy	Population with access to electricity.
S4	Comfort	Population with access to natural gas.
S5	Health	Population with access to fresh water network.
S6	Social justice	Population with legally hired labor with all the social benefits.
S7	Formal education and instruction	Population with tertiary education or college degree.
S8	Acceptance of bioenergy projects	Percentage of public acceptance of bioenergy related projects.
I1	Organizational capacity	Population that belongs to some kind of association or cooperative.
I2	Government efficiency	Population that trusts the local government.
I3	Citizen participation	Population that has taken part in any local government decision.
I4	Political stability	Population who is aware of the political proposal of the government.
I5	Link and integration	Population linked to producers of another area of the Valley (exchanges, leases, membership of cooperatives).
I6	Regulatory quality	Population who is aware of the difficulties to develop its activity in the private sector.
I7	Government support	Number of municipal administrations which support renewable energies projects (logistical, regulatory and financial).
I8	Cooperation	Population who is willing to organize.
E1	Economical stability	Population who observes that the yield of their crops is stable throughout the productive cycles.
E2	Economical solvency	Population with the capacity to save money.
E3	Government foment of the local economy	Percentage of the population who thinks the productive activity infrastructure is enough (roads, installations).
E4	Popular conformity	Population who considers themselves in a good economic situation.
E5	Capitalization	Population who pays a lease.
E6	Access to assets	Population who must lease tools and machinery in order to produce.
E7	Rural economy strength	Population who depends on subsidies and state financing.
E8	Energetic dependency	Percentage of expenses on fossil fuels during the productive cycle.

Table 5. Criteria for assessing the sustainability of bioenergy systems according to the survey to experts

Criteria	Description
Energy balance	Energy balance obtained versus the energy consumed in the bioenergetic chain. It also implies efficient use of the energy by using less fossil fuels and transportation.
GHG balance	CO_2 emissions realized during the bioenergy chain and possible compensation level from the absorption of the biomass during the cycle. Balance close to neutral. Other GHG emissions are CH_4, O_3, NO_2, and H_2O.
Land-use change	Affectation of the deforestation rate. Use of marginal lands or with agricultural ability for the project. Substitution of food crops.
Use efficient of natural resources	Includes cascade use of the biomass, management plans that respect the natural cycles of the resources (the harvest rate must not be superior to the growth), minimizing the use of resources, raw material y lands.
Respect of the human rights	Levels of respect and promotion of the basic human rights (access to health, education, shelter, work, etc). Existence of discriminations, marginalization, exclusion of minorities, abuse against marginal or vulnerable sectors (aboriginal communities, rural settlements, children and women).
Demand satisfaction	Suitability of the project in satisfying the detected needs in the place where it will be executed. The benefits can be perceived locally before externally.
Food security	Competition over the use of the land for food versus energy. Preservation of the food productions and local economies.
Rights of use and ownership	Displacement of communities or destruction of cultural values associated with the territory. Respect for the individual and collective property, with or without the corresponding legalization.
Democratic participation	Level of participation of the different actors and sectors involved or affected, in the processes of decision making. Level of connection in the opinions and suggestions of the actors at a political and administrative level.
Water management	Depletion of the water reserves, existence of a management plan, supply, treatment and distribution. Requirements for the project.
Compliance with the laws and regulations	Strict compliance of the ordinances dispositions regulations and laws of each area of the chain (labor, production related, commerce related etc) on a local, regional and national level.
Ecosystems preservation	Preservation of the connectivity, their biodiversity, must not decrease the density of the locally used multi-purpose species. Must not introduce genetically modified organisms. The quality of the landscape does not seem modified or irreversibly impacted.
Biodiversity protection	Preservation of the vulnerable, rare, threatened or endangered native species. Sustenance of the ecosystem's biodiverse genetic pool by protecting the natural habitat of the wild fauna and flora. If exotic species are introduced, there must be a strict monitor and control system.
Soil protection	Type of practices used (traditional conservationist, etc) and manner in which the structure, fertility, organic material and water storage capacity are conserved. Resting period and rotation of the pastures, green coverage, etc. to avoid erosion and soil wear and tear.
Cost efficiency of the project	Potential benefits of the project and costs and impacts associated with it. Make efficient use of the resources and minimize the impacts. Costs of operation and functioning. Costs of installation.
Economic impact in the region	Sectors and people affected in their economy (due to savings of the use of fossil fuels, decrease in electrical fees or of other fuels like biomass from the projects). The activity can contribute to the strengthening and diversification of its economy.
Local contamination	Chemical dependency, pest control, fertilizers, that might affect the water, soil, air, vegetation or man. Pest control in the water, possible eutrophication of the water bodies due to phosphoresced emission, respiratory and skin diseases, heavy metal accumulation, local residue burning. Includes different GHG emissions like nitrogen emissions and persistent organic polluters, SOx, CO, NOx and particle emissions.
Cultural acceptance	Identification of the cultural, traditional and belief models that might difficult the development of the projects and the population's level of commitment.

Table 5. (Continued)

Criteria	Description
Local energetic security	The security of the energetic supply can be assured with the diversification of the fuel sources that can be used in the area, and the decrease in the use of fossil fuels that are subject to price variation. Access of the local population to the energy.
Job generation	Job generation with fair conditions, training and teachings for the local labor and strengthening of the local abilities.
Short, medium and long term planning	Existence of a planning for the use and management of the resources, of the medium and long term activity development, of the monitoring of the indicators that allow for an adequate verification of the project.
Commerce and income distribution	Development of a fair commerce and a equitable distribution of income to all the links of the bioenergy chain, for which there should be regulation mechanisms, monitoring and government and other possible participating organism control.
Reliability on the proposal	Transparency, public accessibility, level of divulgation and openness accounting mechanisms, ensured initial capital, feedback mechanisms, externality generation.
External input dependency	Measured as production supplies, subsidies and primes for the development of the activity, parts for the correct technological functioning of the proposal, etc.
Residue management	Minimization in the generation of residues, appropriate residue management and disposal, residue recycle when available, biomass combustion ash recycle, added value to the waste. Generation of other solid and liquid residues.
Persistence mechanisms	Will for a government support whether political, financial or of infrastructure. Possible certification mechanisms. Potential sources of financing for the project, available successful experiences. Compatibility with other beneficial alternatives for the region.
Working conditions	Retribution with a fair minimum wage, healthy work journal, prohibition of child labor, payment of all the social benefits and legal covers, protection equipment for the development of the activities, quality of the place of work.
Technology development	Level of knowledge and testing of the knowledge, whether locally or globally. Commercial development phase. Appliance and experience at different scales and regions. Complexity of installation and functioning.
Organization and connection	Links between actors and sectors. Creation or strengthening of co-operatives, associations. Supplies and other aspects of the commercial chain contracts and agreements.
Communication and information	Communication mechanisms and information exchanges among all the links of the chain, as well as conflict resolving and dialogue mechanisms. Free access to the information and generated records and public operations mechanisms.

bioenergy systems could provide. The institutional aspect would probably be slightly modified, and in terms of the "organization" rather than the "institution" per se due to the fact that the proposed projects imply exchanges and interconnections that might result favorable to the producers in other aspects as well. This is an almost instrinsical characteristic of the bioenergy projects, as long as the resource dispersion, the low energy density and the generation seasonality (common factors in the biomass resources) promote regional measures rather than local ones.

The proposals for the use of biomass resources that aren't currently used, as well as the resource management, can noticeably contribute to the conservation of natural capital, decrease in CO_2 emissions (the sustainable use of biomass, respecting the resource renewal rate, results in neutral emissions of CO_2), improvement of the hygiene and environmental sanitary aspects (urban solid waste disposal in a regional sanitary landfill, decreasing impacts and pollution), as well as the recovery of currently wasted energy. This would also impact the energy fees from fossil fuel consumption, with a level of economic saving, which could contribute to a solution to some unsatisfied basic demands: continuous provision of cooking and heating fuel, more homes connected to the power line, supplying of the public wellbeing places (schools, hospitals, clinics, etc.), among others. Correctly planned and executed, this could imply healthy and stable jobs. Regionally, the logistics of the use of biomass requires a certain level of organization and task distribution, and the management needs the local government's support and cooperation agreements because the regulations, financing and other aspects must be resolved at this level in order to give the initiatives strength. Although the local population does not consider the institutional aspects as important, these can result in a real obstacle for these projects. On the other hand, the option of some technological applications over others will be necessary for the successful development of bioenergy: improved and more efficient cooking stoves, technology with fewer emissions and manufactured locally or nationally, and implementation of more efficient boilers for the use of biomass during the productive cycle, among others.

According to the previous appreciations, the bioenergy systems could generally elevate the sustainability levels of the Lerma Valley but, how should be the bioenergy systems so those changes occur? What aspects should be taken into account for evaluating if they are sustainable within themselves?

3.3. Analysis of Bioenergy Systems and Main Impacts on the Level of Local Sustainability

i) The Sustainability Evaluation Criteria

An international survey performed by experts from all over the world (Manrique et al., 2010 a) shows the main criteria (understood as a group of conditions by which an object is evaluated for given dimensions, Gnansounou et al., 2007) with which the bioenergy systems of the Lerma Valley should be evaluated. The Table 5 shows these criteria in a hierarchy order assigned by the consulted experts. Respond to three analysis dimensions: environmental, social and economic, which are considered the pillars of sustainability, in the origins of this concept. In addition, these dimensions are treated internationally in biomass sustainability's policies and programs, as well as in international certification schemes

(Lewandowski and Faaij, 2006; Gnansounou et al., 2007; UNEP, 2009). It is impossible then, to achieve an international understanding and participation if these dimensions weren't considered.

The results of the survey reveal that within the environmental dimension (Figure 3, top), the "energetic balance" carries the largest weight (17% of relative importance), one of the most studied and discussed aspects in the international field of biomass. There are some differences in energetic and balance efficiencies among the currently used energy crops that depend upon the considered variables and the limits or frontiers of the systems being evaluated. According to which components are being evaluated, the results are very different (Pimentel and Patzek, 2005; Pimentel et al., 2007; Goldemberg, 2007; Rajagopal et al., 2007; Laurence, 2007; Campbell et al., 2009). The second important aspect within the environmental dimension is the "GHG emissions balance" (15% of relative importance) which is also subject to great controversy, mainly associated with the incorporation of aspects like land-use change generated by the expansion of the energy crops. Gibbs et al. (2008) point out that the liquid biofuels have the capacity to provide significant reductions in GHG emissions but substituting the fossil fuels, but for this to be a reality depends on how and where those biofuels are produced (Hill et al., 2006, Alder et al., 2007; Gibbs et al., 2008). Some authors argue that the biofuels made from food crops will involve deforestation, resulting in a small reduction of GHG emissions through the release of carbon (Crutzen et al., 2007; Righelato and Spracklen, 2007; Hill et al., 2006; Koh and Ghazoul, 2008). Worst case scenario, for example, the GHG emissions from biofuel production can be higher than from an equal amount of fossil fuels (Delucchi, 2006; Farrell et al., 2006), and particularly when considering the conversion of the natural ecosystems to biofuel plantations, outshining the greater benefits of the bioenergy systems (Righelato y Spracklen, 2007; Fargione et al., 2008; Searchinger et al., 2008). The GHG emissions in the biofuel production will be even higher under poor management conditions –when the soil's organic matter is lost during production, or when too much nitrogen is used as fertilizer (Crutzen et al., 2007; Smeets et al., 2007). These two criteria are the main aspects internationally evaluated in any type of bioenergy system.

Significant aspects are those related to resource management like "water management", "efficiency in the use of natural resources", "land-use change" – mainly related to the possibility of realizing energy crops for biofuels, which is not considered an adequate choice for the area -. "Species and ecosystems protection", will probably have a greater influence when wood resources are considered, since the use of residual resources might not have an impact on species or ecosystems. These residual resources might possibly have a high score in criteria like "local pollution" and "residue management" or including "soil protection". The more or less weight obtained for each criteria focuses on the dimension of the main aspects considered internationally, bur does not exclude that all those criteria must be considered since they all have certain "weight" or level of importance- on another hand, it will depend on the observation of the local context, where the relative importance of the criteria may vary.

Regarding the social dimension, the criteria acquire the relative importance observed in Figure 3 (center). The main social criteria are: "social participation", "communication and information" (also linked to the promotion of participation), "compliance of the current laws and regulations (related to justice and activity control aspects) and "respect for the human rights". The "satisfaction for the local demands" criterion presents with a high relative

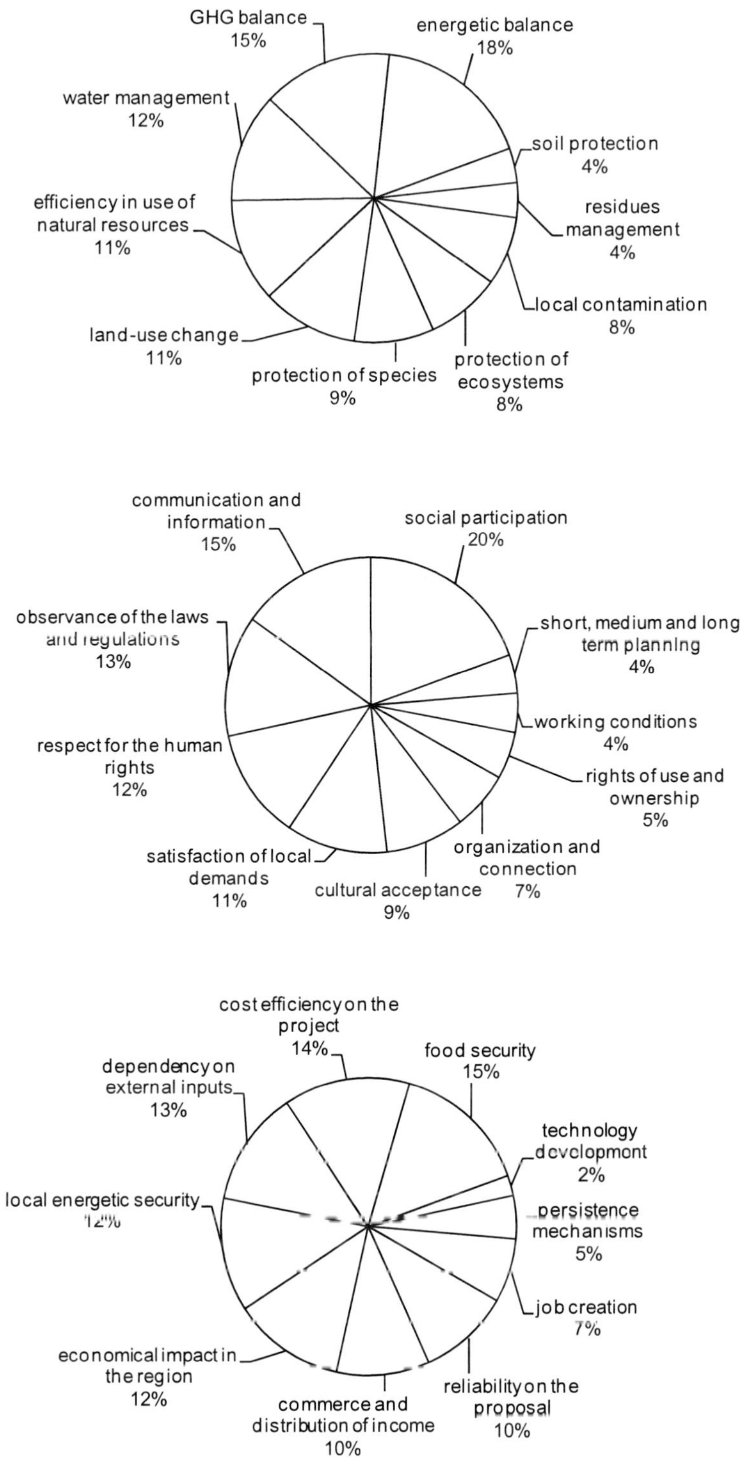

Figure 3. Allocation of the relative importance or considered weight of each criteria of the environmental dimension (top), the social dimension (center), and the economic dimension (bottom).

importance and this one alone can act in some cases as an excluder in some bioenergy proposals. The "cultural acceptance" seeks to incorporate some of the human dimensions that should be taken into account in the evaluation for these types of systems. Domac et al (2005) point out that the essence of the sustainability of bioenergy projects from a social point of view is how society perceives these projects, and how different societies benefit from this activity. Possibly some aspects like "carbon emissions", for example, are added benefits for the local communities, which are more worried about job positions or access to energy. If these aspects, along with others such as "respect for the human rights", are contemplated, they will allow for a larger social cohesion and stability, favoring the permanency of the project through time. As can be observed, although with a more or less importance within the dimension, aspects like "organization and connection", "rights to use and ownership", "labor conditions", and "short, medium and long term planning", are also unavoidable subjects in terms of a more sustainable "regional system". Probably, looking at these dimensions as limiting for the system to show a higher level of sustainability will stop the process of implementation and will demand reassigning the efforts so these aspects may be incorporated in favorable schemes.

Finally, the weight distribution of the different criteria within the economic dimension is observable in Figure 3 (bottom). The "food security" has been marked as one of the most important criterion for the area – in other cases, this criterion was assigned to the social dimension and in some cases, even to environmental dimension – from the international expert's point of view (Koh and Ghazoul, 2008; Von Braun, 2008). The use of crops for the generation of energy, causing inflation and rise of basic food prices, instead of being destined for the satisfaction of the world's population's hunger, is pointed out as unsustainable (Zidansek et al., 2009). Another important criterion in this dimension is related to the costs ("project's cost-efficiency and "external input dependency") and the contribution the proposed systems could make to the regional economy ("economic impact in the region"). There is an incorporation of another criterion that is related to the technology to be used ("technology development"), although it is a partial view in the total technological analysis that must be done for each case for only the development phase in which it stands is considered.

ii) Bioenergy Systems and Sustainability Level

In order to realize the sustainability level analysis for the bioenergy systems proposed for the Lerma Valley, an international survey was created on a spreadsheet where the accounted and ordered according to the mentioned degree of appreciation and importance criteria and were included in a Multi-Criteria Assessment matrix (MCA). The MCA can be defined as a formal approximation that seeks to have explicit multiple criteria accounts to help individuals and groups explore the decisions and solutions that worry them (Mendoza and Martin, 2006; Buchholz et al., 2009 b).

It not only allows to include different aspects of one subject in the analysis of a problem, but also to incorporate different points of view and social actors (Munda, 2003; 2004). The tools based on MCA have shown a high potential in guiding the actors towards finding or achieving an agreement on sustainable solutions in forest management, natural resources and renewable energy management (Kontos et al., 2005; Mendoza and Prabu, 2005; Giampietro et al., 2006; Gamboa and Munda, 2007; Geneletti, 2007; Cavallaro, 2009). There are successful experiences in bioenergy systems where MCA was applied (Begic and Afgan, 2007; Shi et al., 2008; Karagiannidis and Perkoulidis, 2009). This tool has also been considered valid for the evaluation of national and international environmental policies (Greening and Bernow, 2004; Solomon and Hughey, 2007).

This type of Multi-Criteria Assesment begins with the idea that in a real problem, with a very high complexity, there is not one solution that simultaneously optimizes all the criteria, therefore it is necessary to reach a compromising solution between the different values and interests, i.e., a multicriteria solution. The final objective of the MCA techniques is to investigate a number of alternatives under multiple criteria and conflicts. According to this, it is possible to achieve solutions of compromise and alternative hierarchy depending on the degree of importance (Karlsson et al., 2009).

The three use-conversion-application alternatives (for each selected biomass resource) that were defined were incorporated in the MCA for their comparative evaluation in the following bioenergy system based on: option A, USW (sanitary landfill biogas); option B, agricultural residue; option C, biomass woody of different species of the genus Acacia.

These three alternatives were compared with two other energy alternatives: option D, use of lignocelluloses for the creation of second generation biofuels (cellulosic bioethanol) and option E, current scenario of use of fossil fuels (mainly natural gas and gasoil). The option D was incorporated because the current importance of biofuels on a global level, and the controversies generated provincially and nationally. The option E was necessary to evaluate with identical sustainability criteria the energy generation based on fossil fuels.

The expert's participation consisted in assigning a score for every option regarding each criterion. This is, to estimate the impact of the crossings: criterion-bioenergy option, with a 0-100 point scale. The survey looks forward to incorporate different perspectives of importance for each option according to the previously defined criteria. This process could be repeated with other involved actors.

The MCA resolution method used was the Weighed Lineal Addition (WLA), also known as the SMART method (Simple Multi-Attribute Rating Technique), based on a matrix methodology. This method involves assigning a value to different possible options, which are weighed with importance values previously assigned to a series of criteria.

The WLA provides a unique value that allows for comparison and order the options. Despite its simplicity this method provides reliable values (Hajkowicz, 2007) and it is one of the most applied methods (Pohekar and Ramachandran, 2004). The MCA results were processed following the sustainability level scale analysis: very low (0-25%), low (25-50%), medium (50-75%) and high (over 75%). The proposed evaluation scale allowed the comparative observation of the sustainability level of each of the five studied alternatives. The alternative acceptance threshold can be discussed and defined in each case, it depending on the level of impact that is considered "allowed" in each site. In Manrique et al (2011) was defined that the alternatives would be valid for application if the sustainability level was 50% or more (medium or high sustainability level). This decision doesn't inhibit that each criteria in which the option shows a low level, should be treated in an individual manner.

Three possible scenarios are shown in order to analyze the strong and weak points (Figure 4). These scenarios were defined from the most frequent answers from experts in the international survey:

a) The USW (option A) and the wood biomass (option B) result in the system with the highest scores or sustainability levels (with a predominance of the first over the second). In both cases, they are considered as highly sustainable alternatives. Options D and E are options that could be discarded or revised in those criteria in which they have a low score. Option C (agricultural residues), with a medium level of sustainability is a viable alternative. Application of option A would positively impact the region reducing the pollution of soil, air and water. Also this system would use resources and energy that otherwise be wasted.
b) The woody biomass (option B) as an energy source is the most favored, mainly because it relies on a resource that is used traditionally in the region and has a high heating value. This option is almost 100% sustainable. It is the "most optimistic" option. Option C is located in second place followed by option A (USW). All three alternatives present medium to high levels of sustainability. Option D (cellulosic bioethanol) resulted as a feasible system for implementation in the area. The fossil fuel option resulted with the lowest scores.
c) The option C has the highest score but with values similar to the other two bioenergy options. The sustainability levels obtained for options D and E are low. Once again, the woody biomass is a good choice, arguing the qualities of this type of residues to be used in combustion processes (option B), along with agricultural residues (option C). This scenario is the most "pessimist" because all the options have low score and the proposed systems have a "medium sustainability" qualification. Options D and E reach low levels of sustainability.

Although option D corresponds to a possible bioenergy system, there is not enough knowledge on the energy conversion technology and will have associate high costs. This alternative, although with a low level of performance in many of the MCA criteria, could become and adequate alternative in the future.

Not one of the criteria reach a 100% performance for any of the bioenergy systems, which points out the complexity and uncertainty involved in any system or real entity analysis. Clearly, is important the performance of the options in some criteria. It is the same to obtain a low score in a criterion like "energetic balance" (higher hierarchically) that in one such as "communication and information" (lower hierarchically). As the criteria are arranged according to their relative importance (rating given by the experts), each criterion has a different performance depending of the hierarchy level. According to that, the score may or may not be restrictive wich should be agreed upon by the actors involved in each case.

The performance of each option in each criterion, averaged for the three scenarios, is shown on Figure 5. In the consulted expert's opinion, none of the options is 100% sustainable. Except for few exceptions, 100% was not allocated to the individual analysis of the criteria performance. This shows certain "carefulness" or the expert's recognition of the multiple factors that influence each reality.

The proposed bioenergy systems show a level of sustainability that ranges from medium to high. Although at first it is considered that the three options or strategies could be developed simultaneously, this tool allows the detection of aspects that are influencing the level of sustainability of each alternative. It is possible to decide by the implementation of the most favorable alternative in the short term, and incorporate other alternatives in a strategic medium and long term plan.

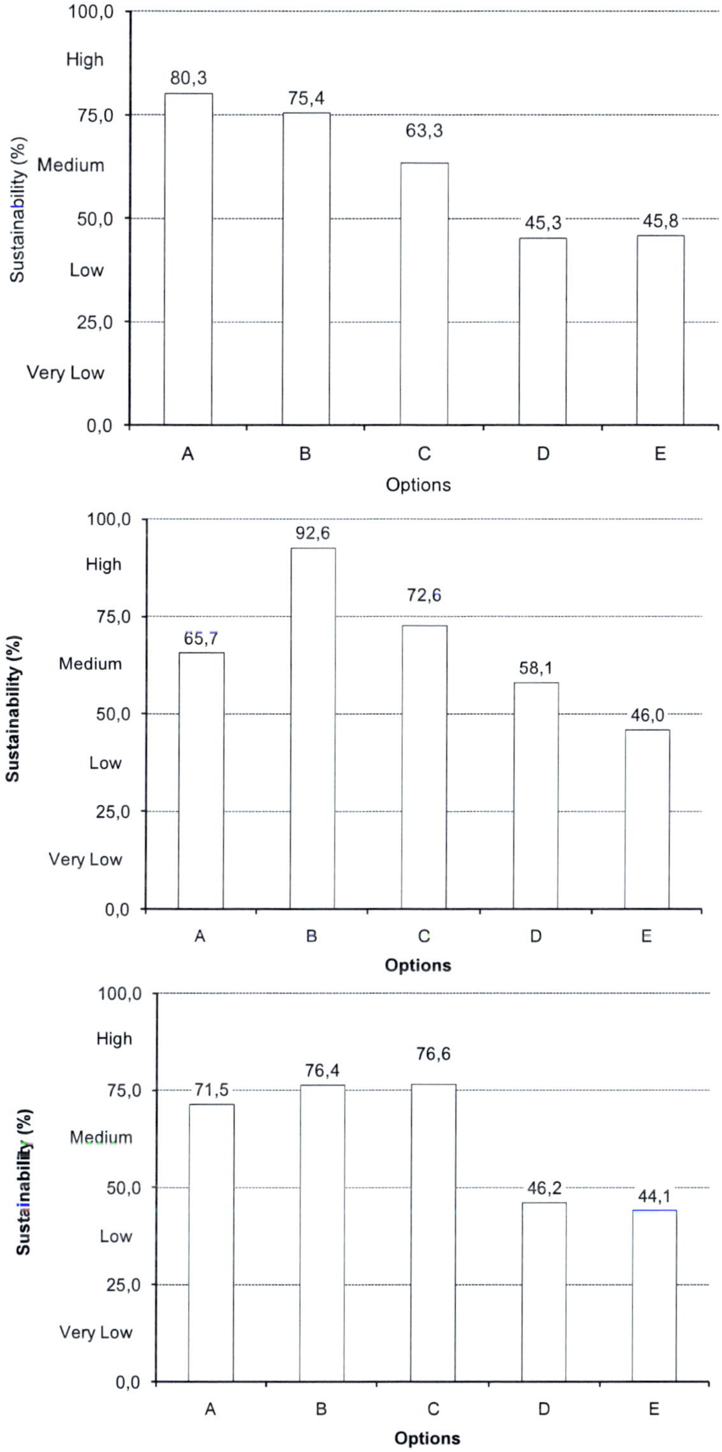

Figure 4. Three most common scenarios obtained in the sustainability assessment of the energy systems: (top) USW system; (center) woody biomass of Acacia spp.system; (bottom) agricultural residue biomass system. Options: (A) generation of electric energy from USW; (B) thermal generation from woody biomass (Acacia spp.); (C) thermal generation from agricultural residues; (D) liquid biofuels from woody biomass; (E) thermal and electrical generation from fossil fuels.

For option D and E the main criteria that show the lower values are:

- energetic balance: there is a lot of discussion about this, but because it is a technology in evaluation phase and not commercial it could imply a high energy demand;
- land-use change: this alternative could promote greater deforestation;
- satisfaction of local demands: is low possibly because these types of fuel are usually destined for export (at least in the case of developing countries);
- economic impact (low), improvement in life conditions and access to energy in the region (depending on the production's destination);
- low cultural acceptance: due to some impacts associated with biodiesel export soybean monocultives;
- high input dependency, among others.

For the option E lower values are observed in: the GHG emissions balance (absolutely negative); the low efficiency of the use of the resource (includes a lower use than the generation rate); the way the activity takes places in terms of low democratic participation (can be discussed according to each region); how the right of use and property of the land is affected (displacement of local communities or destruction of cultural values); alteration of the ecosystem and species (especially when the activity takes place in protected areas or of high ecological importance); energetic insecurity (due to fluctuating prices resource's dependency and with a limited reserve perspective); monopoly development of the activity, among others.

The options B and A are the "most sustainable" options. Both alternatives could be improved, deeply researched and implemented with a greater probability of achieving major positive than negative impacts. In the case of the shrubland (option B) the energetic balance, the GHG emissions and the land-use change, have a high values because the use of this woody biomass does not imply modify the ecosystem or the current use of land. If the biomass is harvesting by a manual system there are not machine movements and the energetic demand will be low.

As long as the natural productive rate of the used species is respected, the GHG emissions will be low and close to zero. Regarding the food and energy security, as it has been analyzed before, the use of wood biomass is a regular practice for the Valley inhabitants.

If it will be managed, it would not affect the land for the production of food. These are native species that are adapted to the area, capable of survival in diverse conditions; therefore, the use of this species would not demand irrigation – scarce resource in the area. It would allow keeping the vegetative coverage and protecting the soil.

The application of this alternative would not imply a negative impact on the environment, local pollution, waste generating at a scale that would become adverse.

The technology development criterion shows a good performance because the technology available for combustion system is the simple implementation. The traditional use of the wood respects the rights of use and property. These types of project could be support by for forest management certification mechanisms, carbon credits or other mechanism.

In the case of the USW (A), the criteria in which this option shows good performance, assuming that will be the regional sanitary landfill are:

- efficiency in the use of natural resources and residue management: the amount energy of previously wasted fractions could be used (they should be store in a place specially fitted);
- local contamination would be reduced with a positive impact for the natural and social environments;
- protection of species, ecosystems and soil: probably, the impact the USW caused over species, ecosystems and soil would decrease;
- compliance with the laws and regulations; there is a short/medium and long term plan for the development of this project within the province. This has a positive influence in the persistence mechanisms criterion as well as is well managed technical and administratively. This project doesn't alter of the land use, only a small space for the disposition of the residues;
- food security would not be affected nor would it threaten the current productive practices, which it is an alternative with a high cultural acceptance;
- government planning, the jobs would enjoy all the minimum social benefits (labor conditions).
- development of the technology shows a high performance because there is a lot of experience in the anaerobic digestion process from sanitary landfills and electrical generation.
- GHG emissions may be argued because if considered from the use stage, tones of annual methane emissions into the atmosphere will be avoided. In this sense, there will be a positive balance of GHG emissions (more than what is emitted is sequestered or eliminated). If the analysis incorporates the Life Cycle Assessment (LCA) of the elements that appear as landfill residues, the balance will probably be negative.
- respect for human rights got a high score possibly because the use of bioenergy from USW implies cleaner and healthier environment for the Valley.

Among the criteria for which this option (USW) shows a low (or medium) performance are:

- energetic balance, because energy is required for the transportation, treatment and redistribution of this residue (this is without taking into account the energy invested in the fabrication process of each of these elements);
- satisfaction of local demands and job generation, as well as the local energetic security, perhaps because the energy benefits are limited due to the low potential of the resource;
- economic impact in the region as well as the cost-efficiency of the project are low if the project not taken into account carbon credit mechanisms (could also be economically profitable if there were laws of specific subsidies or incentives).

In the case of the agricultural resources (option C), some of the criteria in which this option shows low performance are: efficiency in the use of natural resources, water management, soil management, soil protection, local pollution, energetic balance, GHG

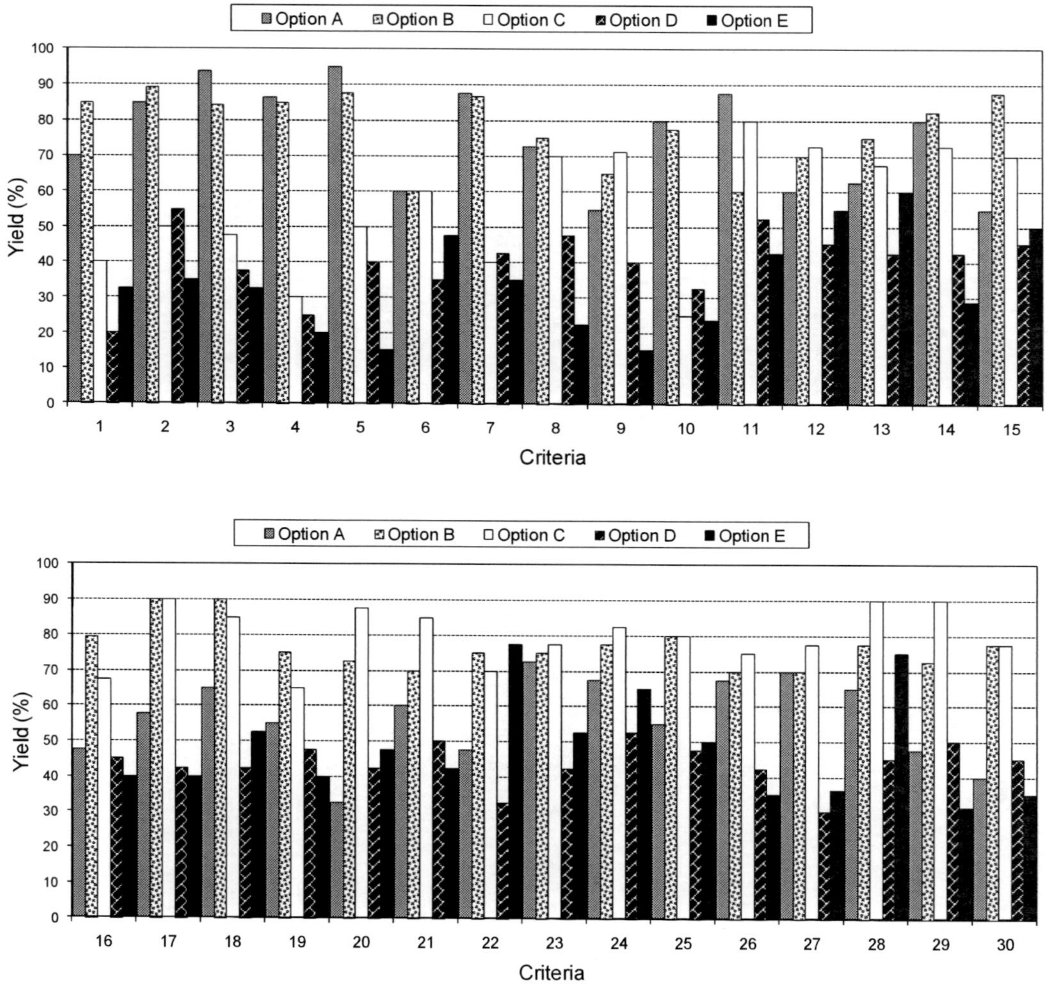

Figure 5. Yield of each criterion for every option (see criteria details in Table 5). Option A: generation of electrical energy from USW; option B: thermal generation with woody biomass (Acacia spp.); option C: thermal generation with agricultural residues; option D: liquid biofuels from woody biomass. Top: criteria of 1 to 15. Bottom: criteria of 16 to 30.

emissions, land-use change, local demands satisfaction. Although has some aspects positive: resource potential, labor, technology available, favorable legislation, and others. These results show that any energy use proposal must contemplate the whole system, because it could be good in some aspects but in other aspects it may cause serious impacts and alter the systems tendency or analysis unit in a negative way.

Each of the proposed bioenergy options will probably impact the Valley's current situation, but if the application is isolated, this impact will be restricted. It is important to combine several of bioenergy proposals with a long term integral policy that will not only support these types of initiatives, but will also boost their effects and generate additional benefits.

All the considered options show "some level of sustainability", including the use of fossil fuels which logically provide some recognized benefits (great technological development which can be used for centralized and decentralized applications, automotion, heat), and will continue to be used in the region and the world, at least in the short and medium term. On the other hand, the "acceptance threshold" of the alternatives can be discussed. If the sustainability level is "low" or "very low", the community could look for new alternatives.

The A, B and C systems could be optimized in several criteria, for example, obtaining the land property of the communities, promoting the consolidation of associations or cooperatives, changing the farming or cultivation practices, respecting the ecological corridors, among many other aspects.

The developed methodology and the results shown in this chapter appear to be a useful contribution in bioenergy production studies, detecting and quantifying critical aspects that must be evaluated. These results therefore are of great importance not only locally and regionally but also globally, and might result of particular interest (Rametsteiner and Simula, 2003) for: i) the industry and commerce, as an environmental marketing and market access tool, and as tool for controlling the production line (origin and quality of the feedstocks, products or services); ii) buyers and consumers, providing the information on the impact of the products or services they are consuming, improving their trust on them; iii) producers, as a tool for accessing the markets or increase of their advantages, allowing for the differentiation of the products; iv) governments, as a political tool for promoting the sustainable management and the "sustainable" consumer patterns, providing information for the political advice.

In Argentina, bioenergy has a low participation in the national energy matrix (approximately 2%), so that there is a biomass promotion through differents laws and incentives. However, the quality standards have not yet been developed to allow controlling, monitoring and certifying that the "bioenergy" does not produce adverse impacts. This work could be use in that direction.

CONCLUSION

The energy sustainability demands consideration and satisfaction of the society's needs, not only of physical survival but also economics, social, political, cultural and environmental needs for a dignified life. This concept should be evaluated for particular alternatives and contexts, defining the criteria for analysis. If the criteria are defined properly and in a participatory way, it will obtain an approximation to the level of sustainability of the bioenergy systems proposed from multiple perspectives and dimensions.

The developed methodology and the results are a useful contribution to the advancement of knowledge of bioenergy systems, detecting and quantifying critical aspects which should be evaluated, and which could be included in quality standards.

The explored bioenergy systems show sustainability levels that range from medium to high, which indicate systems that, when implemented in the Lerma Valley, could clearly imply greater benefits than harm. Even more, the detection of dimensions and criteria with some level of problem (or low performance) will allow worked upon them and improved. The three studied systems could be implemented simultaneously with a positive impact in the local sustainability.

References

Afgan, N.H and Carvalho, M.G. (2008). Sustainability assessment of a hybrid energy system. *Energy Policy* 36, 2903– 2910.

Alburquerque, F. (2004). Desarrollo económico local y descentralización en América Latina. *Revista de la CEPAL* 82, 157–171.

Alder, P.R., Grosso, S.J. and Parton, W.J. (2007). Life cycle assessment of net greenhouse-gas flux for bioenergy cropping systems. *Ecological Appliccations* 17, 675–691.

Begic, F. and Afgan, N.H. (2007). Sustainability assessment tool for the decision making in selection of energy system—Bosnian case. *Energy* 32, 1979–1985.

Bell, S. and Morse, S. (2005). Holism and understanding sustainability. *Systemic practice and action research* 18 (4), 409-425.

Bermann, C., Aedo, M.P. and Larraín, S. (2004). Desafíos para la sustentabilidad energética en el Cono Sur. Programa Cono Sur Sustentable. Fundación Heinrich Boll.

Bertinat, P. and Salerno, J. (2006). Un modelo energético en apuros. Alternativas para la sustentabilidad energética en Argentina. *Fundación* Heinrich Boll.

Bertinat, P. (2007). Agrocombustibles. Argentina frente a una nueva encrucijada. *Fundación Heinrich Böll. Inercia*. Argentina. 28 p.

Bindrabán, P.S., Bulte, E.H. and Conijn, S.G. (2009). Can large-scale biofuels production be sustainable by 2020? *Short Communication. Agricultural Systems* 101, 197–199.

Binimelis, R., Pengue, W. and Monterroso, I. (2009). ''Transgenic treadmill'': Responses to the emergence and spread of glyphosate-resistant johnsongrass in Argentina. *Geoforum* 40, 623–633.

Bossel, H. (1999). Indicators for Sustainable Development: theory, method, applications. IISD Publications Centre. 124 p, ISBN 1-895536-13-8.

Buchholz, T., Rametsteiner, E., Volk T.A. and Luzadis, V.A. (2009a). Multi Criteria Analysis for bioenergy systems assessments. *Energy Policy* 37, 484–495.

Buchholz, T., Luzadis, V.A. and Volk, T.A. (2009b). Sustainability criteria for bioenergy systems: results from an expert survey. *Journal of Cleaner Production* 17 (1), 86–98.

Bush, S.R. (2008). The social science of sustainable bioenergy production in Southeast Asia. *Biofuels, Bioproducts and Biorefining* 2,126–132.

Cabrera, A.L. (1994). Regiones Fitogeográficas Argentinas. Enciclopedia Argentina de Agricultura y Jardinería. Ed. Acme Saci, Bs. AS.

CAER (Cámara Argentina de Energías Renovables). (2008).Outlook for the Argentine biodiesel industry. Buenos Aires. /http://www.argentinarenovables.org/informes_estudios _ensayos.phpS.

Campbell, A., Dickson, B., Kapos, V., Miles, L. and Scharlemann, J. (2009). Biofuels and biodiversity: an assessment of the impacts of biofuels on biodiversity and the appropriate policy responses. Climate Change: global risks, challenges and decisions. *Earth and Environmental Science* 6.

Carvahlo, R. (1999).The Amazon: towards the ''soy bean cycle''. *Amazonia papers* 2 (1). CESPA.

Cavallaro, F. (2009). Multi-criteria decision aid to assess concentrated solar thermal technologies. *Renewable Energy* 34, 1678–1685.

CESPA (Centro de Estudios de la Situación y Perspectivas de Argentina). (2007). Biocombustibles: expansión de una industria naciente y posibilidades para Argentina. Universidad de Buenos Aires, Argentina. /http://www.econ.uba.ar/ www /institutos/economia/CESPA/index.htms.

Costanza, R., D'Arge, R., De Groot, R., Farber, S., Grasso, M., Hannon, B., Limburg, K., Naeem, S., O'Neill, R.V., Paruelo, J., Raskin, R.G., Sutton, P., Van den Belt, M. (1997). The value of the world's ecosystem services and natural capital. *Nature* 387, 253–260.

Crutzen, P.J., Mosier, A.R., Smith, K.A. and Winiwarter, W. (2007). N2O release from agrobiofuel production negates global warming reduction by replacing fossil fuels. Atmospheric Chemistry and Physics (ACP) *Journal* 7, 11191–11205.

De Groot, R.S., Wilson, M.A. and Boumans, R.M.J. (2002). A typology for the classification, description and valuation of ecosystem functions, goods and services. Special Issue: The dynamics and value of Ecosystem Services: integrating economic and ecological perspectives. *Ecological Economics* 41, 393–408.

Delucchi, M.A. (2006). Lifecycle Analyses of Biofuels. Draft Report. UCD-ITS-RR-06-08. Institute of Transportation Studies. University of California, Davis.

Demirbas, A.H. and Demirbas, I. (2007). Importance of rural bioenergy for developing countries. *Energy Conversion and Management* 48, 2386–2398.

Demirbas, A. (2009). Political, economic and environmental impacts of biofuels: A review. *Applied Energy* 86, 108–117.

Dincer, I. and Rosen, M.A. (1998). A worldwide perspective on energy, environment and sustainable development. *International Journal of Energy Research* 22, 1305-1321.

Domac, J., Richards, K. and Risovicc, S. (2005). Socio-economic drivers in implementing bioenergy projects. *Biomass and Bioenergy* 28, 97–106.

Domínguez-Faus, R., Powers, S., Burken, J. and Alvarez, P.J. (2009). The Water Footprint of biofuels: a drink or drive issue?. *Environmental Science and Technology* 43, 3005–3010.

Dros, J.M. (2004). Managing the soy boom: two scenarios of soy production expansion in South America. *AI D Environment*, Amsterdam.

Duke, S.O. and Powles, S.B. (2008).Glyphosate:a once in a century herbicide. *Pest Management Science* 64, 319–325.

Elghali, L., Clift, R., Sinclair, P., Panoutsou, C. andBauen, A. (2007). Developing a sustainability framework for the assessment of bioenergy systems. *Energy Policy* 35, 6075–6083.

EUBIA (European Biomass Industry Association). www.eubia.org.

Evans, A., Strezov, V. and Evans, T. (2009). Assessment of sustainability indicators for renewable energy technologies. Renewable and Sustainable *Energy Reviews* 13, 1082–1088.

FAO (Food and Agricultural Organization). (2008). Food and Agricultural Organization Statistics: Production. FAO, Rome. /http://faostat.fao.org/default.aspxS.

FAO (Food and Agricultural Organization). (2009). Análisis del balance de energía derivada de biomasa en Argentina - WISDOM Argentina-Informe Final. Departamento Forestal Dendroenergía. *Financiado por el Proyecto* TCP/ARG/3103. 118 p.

Fargione, J., Hill, J., Tilman, D., Polasky, S. and Hawthorne, P. (2008). Land clearing and the biofuel carbon debt. *Science* 319 (5867), 1235–1238.

Farrell, A.E., Plevin, R.J., Turner, B.T., Jones, A.D., O'Hare, M. and Kammen, D.M. (2006). Ethanol can contribute to energy and environmental goals. *Science* 311, 506–508.

Fitzherbert, E.B., Struebig, M.J., Morel, A., Danielsen, F., Bruhl, C.A., Donald, P.F. and Phalan, B. (2007). How will oil palm expansion affect biodiversity? *Trends in Ecology and Evolution* 23 (10), 538-545.

Fundación Bariloche. (2009). Energías renovables. Diagnóstico, barreras y propuestas. REEEP (Renewable Energy and Energy Eficiency Partnership), Secretaría de Energía de la Nación, y Fundación Bariloche. Argentina. 60 p.

Gamboa, G. and Munda, G. (2007).The problem of wind farm location: a social multi-criteria evaluation framework. *Energy Policy* 35, 1564–1583.

Geneletti, D. (2007). An approach based on spatial multicriteria analysis to map the nature conservation value of agricultural land. *Journal of Environmental Management* 83, 228–235.

GENREN (Programa de Generación de energía eléctrica a partir de fuentes renovables). Argentina. http://energia3.mecon.gov.ar/contenidos/verpagina.php?idpagina=3065.

George, C. (1999). Testing for sustainable development through environmental assessment. *Environmental Impact Assessment Rev*iew 19, 175-200.

Giampietro, M., Mayumi, K. and Munda, G. (2006). Integrated assessment and energy analysis: quality assurance in multi-criteria analysis of sustainability. *Energy* 31, 59–86.

Giarracca, N. and Teubal, M. (2002). El campo argentina en la encrucijada: estrategias y resistencias sociales, ecos en la ciudad. Alianza Editorial, Buenos Aires.

Gibbs, H. K., Johnston, M., Foley, J., Holloway, T., Monfreda, C., Ramankutty, N. and Zaks, D. (2008). Carbon payback times for crop-based biofuel expansion in the tropics: The effects of changing yield and technology. *Environmental Research Letters* 3 (034001): 10pp.

Gnansounou, E., Panichelli, L. and Villegas, J.D. (2008). Sustainable liquid biofuels development fortransport, working paper, *LASEN*, Ref.437.103. EPFL http://bpe.epfl.ch.

Goldemberg, J. and Guardabassi, P. (2009). Are biofuels a feasible option? *Energy Policy* 37,10–14.

Greening, L.A. and Bernow, S. (2004). Design of coordinated energy and environmental policies: use of multi-criteria decision-making. *Energy Policy* 32 (6), 721–735.

Haas, R., Nakicenovic, N., Ajanovic, A., Faber, T., Kranz, L., Muller, A. and Resch, G. (2008). Towards sustainability of energy systems: A primer on how to apply the concept of energy services to identify necessary trends and policies. *Energy Policy* 36, 4012–4021.

Haberl, H., Fischer-Kowalski, M., Krausmann, F., Weisz, H. and Winiwarter, V. (2004). Progress towards sustainability? What the conceptual framework of material and energy flow accounting (MEFA) can offer. *Land Use Policy* 21, 199–213.

Hajkowicz, S. (2007). A comparison of multiple criteria analysis and unaided approaches to environmental decision making. *Environmental Science and Policy* 10, 177–184.

Hammond, G.P. (2004). Towards sustainability: energy efficiency, thermodynamic analysis, and the 'two cultures'. *Energy Policy* 32, 1789–1798.

Hatje, W. and Ruhl, M. (2000). Use of biomass for power- and heat-generation: possibilities and limits. *Ecological Engineering* 16, 41–49.

Hill, J., Nelson, E., Tilman, D., Polasky, S. and Tiffany, D. (2006). Environmental, economic, and energetic costs and benefits of biodiesel and ethanol biofuels. *Proceedings of the National Academy of Science.* U.S.A. 103, 11206.http://www.pnas.org /content/103/30/11206.full.

Holden, E. and Linnerud, K. (2007). The Sustainable Development Area: satisfying basic needs and safeguarding ecological sustainability. *Sustainable Development* 15, 174–187.

Honty, G., and E. Gudynas. (2007). Agrocombustibles y desarrollo sostenible en América Latina y el Caribe. Serie Observatorio del Desarrollo, disponible en http://www.agrocombustibles.org. CLAES. Centro Latino Americano de Ecología Social. Desarrollo, Economía, Ecología y Equidad. América Latina.

INDEC (Instituto Nacional de Estadísticas y Censos). (2011). http://www.Indec.Gov.Ar /Webcenso/Index.Asp.

Joensen, L., Semino, S. and Paul, H. (2005). Argentina: A case study on the impact of genetically engineered soya. How producing RR soya is destroying the food security and sovereignty of Argentina. A report for the Gaia Foundation, London, http://www.econexus.info/publications.htmlS.

Johnston, M. y Holloway, T. (2007). A global comparison of national biodiesel production potentials. *Environmental Science and Technology* 41(23), 7967–7973.

Kaimowitz, D. and Thiele, G. (1999). The effects of structural adjustment on deforestation and forest degradation in lowland Bolivia. *World Development* 27(3), 505–520.

Kalitventzeff, B., Maréchal, F. and Closon, H. (2001). Better solutions for process sustainability through better insight in process energy integration. *Applied Thermal Engineering* 21, 1349-1368.

Karagiannidis, A. and Perkoulidis, G. (2009). A multi-criteria ranking of different technologies for the anaerobic digestion for energy recovery of the organic fraction of municipal solid wastes. *Bioresource Technology* 100, 2355–2360.

Karekezi, S., Lata, K. and Coelho, S.T. (2004). Traditional Biomass Energy. Improving its use and moving to Modern Energy Use. *Thematic Background Paper*. Editing: Secretariat of the International Conference for Renewable Energies, Bonn 2004.

Karlsson, G. (2009). Handbook for conducting technology needs assessment for climate change. Advance document. United Nations Development Programme (UNDP). USA.

Kemmler, A. and Spreng, D. (2007). Energy indicators for tracking sustainability in developing countries. *Energy Policy* 35, 2466–2480.

Kern, F. and Smith, A. (2008). Restructuring energy systems for sustainability? Energy transition policy in the Netherlands. *Energy Policy* 36 (2008) 4093– 4103.

Koh, L.P. and Ghazoul, J. (2008). Biofuels, biodiversity, and people: Understanding the conflicts and finding opportunities. *Review. Biological conservation* 141, 2450–2460.

Koh, L.P. and Wilcove, D.S. (2007). Cashing in palm oil for conservation. *Nature* 448, 993–994.

Koh, L.P. (2007). Potential habitat and biodiversity losses from intensified biodiesel feedstock production. *Conservation Biology* 21, 1373–1375.

Kontos, T.D., Komilis, D.P. and Halvadakis, C.P. (2005). Siting MSW landfills with a spatial multiple criteria analysis methodology. *Waste Management* 25, 818–832.

Kumar, M. and Kumar, P. (2008). Valuation of the ecosystem services: A psycho-cultural perspective. *Ecological Economics* 64, 808– 819.

Lamers, P., McCormick, K. and Hilbert, J.A. (2008). The emerging liquid biofuel market in Argentina: implications for domestic demand and international trade. *Energy Policy* 36,1479–1490.

Lamers, P. (2006). Emerging liquid biofuel markets: A dónde va la Argentina?. *IIEE thesis*: Lund, Sweden.

Laurence, W.F. (2007).Switch to corn promotes Amazon deforestation. Science 318, 1721.

Lewandowski, I. and Faaij, A.P.C. (2006). Steps towards the development of a certification system for sustainable bio-energy trade. *Biomass and Bioenergy* 30, 83–104.

Løken, E. (2007). Use of multicriteria decision analysis methods for energy planning problems. *Renewable and Sustainable Energy Reviews* 11, 1584–1595.

Luke, T.W. (2005). Neither sustainable nor development: reconsidering sustainability in development. *Sustainable Development* 13(4), 228–238.

Lumley, S. and Armstrong, P. (2004). Some of the nineteenth century origins of the Sustainability concept. *Environment, Development and Sustainability* 6, 367–378.

Malthus, T.R. (1798). *An essay on the Principle of Population*. Oxford University Press.

Manrique, S., Franco, J., Núñez, V. and Seghezzo, L. (2008). Potencial energético de biomasa residual de tabaco y ají en el municipio de Coronel Moldes (Salta – Argentina). AVERMA 12 (6): 87- 94. ISSN 0329-5184.

Manrique, S., Franco, J., Núñez, V. and Seghezzo, L. (2009). Estimación de densidad de biomasa aérea en ecosistemas naturales de la Provincia de Salta. AVERMA 13 (6): 37-45. ISSN 0329-5184.

Manrique, S., Franco, J. Núñez, V. and Seghezzo, L. (2010a). Bioenergía en el Valle de Lerma: Evaluación de Sustentabilidad de Alternativas. AVERMA 14 (6): 31-38. ISSN 0329-5184.

Manrique, S., Franco, J., Núñez, V. and Seghezzo, L. (2010b). Perspectiva bioenergética de los residuos sólidos urbanos en el Valle de Lerma. AVERMA 14 (6): 47-54. ISSN 0329-5184.

Manrique, S., Franco, J., Núñez, V. and Seghezzo, L. (2010c). Opciones bioenergéticas y su contribución al desarrollo sustentable del Valle de Lerma. AVERMA 14 (6): 39-46. ISSN 0329-5184.

Manrique, S.; Franco, J.; Núñez, V. and Seghezzo, L. (2011). Propuesta metodológica para la toma de decisiones sobre bioenergía en un contexto complejo y diverso. AVERMA 15 (6): 39-47. ISSN 0329-5184.

Manrique, S. and Franco, J. (2012a). Los cultivos agrícolas del municipio de Coronel Moldes (dpto La Viña, provincia de Salta) y su potencial aporte en el marco del cambio climático. AVERMA (In press).

Manrique, S. and Franco, J. (2012b). Alcances y restricciones de la bioenergía en el municipio de Coronel Moldes (dpto La Viña, provincia de Salta, Argentina). AVERMA (In press).

Mauerhofer, V. (2008). 3-D Sustainability: An approach for priority setting in situation of conflicting interests towards a Sustainable Development. *Ecological economics* 64: 496–506.

May, J.R. and Brennan, D.J. (2006). Sustainability assessment of Australian electricity generation. *Process Safety and Environmental Protection*, 84(B2), 131–142.

Mebratu, D. (1998). Sustainability and sustainable development: historical and conceptual review. *Environmental Impact Assessment Review* 18, 493-520.

Mendoza, G.A. and Martins, H. (2006). Multi-criteria decision analysis in natural resource management: A critical review of methods and new modelling paradigms. *Forest Ecology and Management* 230, 1–22.

Mendoza, G.A. and Prabhu, R. (2003). Qualitative multi-criteria approaches to assessing indicators of sustainable forest resource management. *Forest Ecology and Management* 174, 329–343.

Mendoza, G.A. and Prabhu, R. (2005). Combining participatory modeling and multi-criteria analysis for community-based forest Management. *Forest Ecology and Management* 207,145–156.

Mitcham, C. (1995). The concept of sustainable development: origins and ambivalence. *Technology in Society* 17(3), 311-326.

Mog, J.M. (2004). Struggling with sustainability—a comparative framework for evaluating sustainable development programs. *World Development* 32 (12), 2139–2160.

Mol, A.P.J. (2007). Boundless Biofuels? Between environmental sustainability and vulnerability. *Sociologia Ruralis* 47 (4), 297-315.

Morín, E. and Kern, A.B. (2002). La agonía planetaria. En revista Globalización. Edgar Morin y Anne B. Kern http://rcci.nct/globalizacion/2002/fg270.htm.

Morín, E. (2006). El método I. La naturaleza de la naturaleza. 3ª edición. Madrid, Cátedra.

Munda, G. (2003). Multicriteria Assessment. Internet Enciclopaedia of Ecological Economics. International Society for Ecological Economics. http://www.ecoeco.org /publica/encyc_entries/Mlticritassess.pdf].

Munda, G. (2004). Social Multi-criteria Evaluation: Methodological foundation and operational consequences. *European Journal of Operational Research* 158, 662 677.

Naik, S.N., Goud, V.V., Rout, P.K. y Dalai, A.K. (2009). Production of first and second generation biofuels: A comprehensive review. Renewable and *Sustainable Energy Reviews* 13 (2), 318-345.

Ness, B., Piirsalu, E.U., Anderberg, S. y Olssona, L. (2007). Categorising tools for sustainability assessment. *Ecological economics* 60, 498– 508.

Núñez, V., Menéndez, M. and Moreno, R. (2005). Ordenamiento territorial del Valle de Lerma. CIUNSA (Consejo de Investigaciones de la UNSa). Universidad Nacional de Salta. Salta. Argentina.

Panichelli, L., Dauriat, A. and Gnansounou, E. (2009). Life cycle assessment of soybean-based biodiesel in Argentina for export. *International Journal of Life Cycle Assessment* 14,144–159.

Paz, L.R.L., da Silva, N.F. and Rosa, L.P. (2007). The paradigm of sustainability in the Brazilian energy sector. *Renewable and Sustainable Energy Reviews* 11, 1558–1570.

Phalan, B. (2009). The social and environmental impacts of biofuels in Asia: An overview. *Applied Energy* 86, 21–29.

Phillis, Y.A. and Andriantiatsaholiniaina, A. (2001). Sustainability: an ill-defined concept and its assessment using fuzzy logic. *Ecological Economics* 37, 435–456.

Pimentel, D. and Patzek, T.W. (2005). Ethanol production using corn, switchgrass, and wood; biodiesel, production using soybean and sunflower. *Natural Resources Research* 14, 65-76.

Pimentel, D., Patzek, T. y Cecil, G. (2007). Ethanol production: energy, economic, and environmental losses. *Reviews of Environmental Contamination and Toxicology* 189, 25.

PNUMA (Programa de las Naciones Unidas para el Medio Ambiente). (2004). Informe del Estado del Ambiente - GEO Argentina. Oficina Regional para América Latina y el Caribe (PNUMA-ORPALC) y la Secretaría de Ambiente y Desarrollo Sustentable de la República Argentina. 270 p.

Powles, S.B. (2008). Evolved glyphosate-resistant weeds around the world: lessons to be learnt. *Pest Management Science* 64, 360–365.

Qaim, M. and Traxler,G. (2005). Roundup ready soybeans in Argentina: farm level and aggregate welfare effects. *Agricultural Economics* 31(1),73–86.

Rahman, S.H., Ahmed, A.M., Bauer, A., et al. (2008). Soaring food prices: response to the crisis. Asian Development Bank, Manila, Philippines. <http://www.adb.org/Documents /Papers/soaringfood- prices/soaring-food-prices.pdf>.

Rajagopal, D., Sexton, S.E., Roland-Holst, D. and Zilberman, D. (2007). Challenge of biofuel: filling the tank without emptying the stomach?. *Environmental Research Letter* 2: doi:10.1088/1748-9326/2/4/044004.

Rametsteiner, E. and Simula, M. (2003): Forest certification--an instrument to promote sustainable forest management? *Journal of Environment Management* 67(1), 87-98.

Richardson, J. and Verwijst, T. (2005). Sustainable bioenergy production systems: environmental, operational and social implications. *Biomass and Bioenergy* 28 (2005) 95–96.

Righelato, R. and Spracklen, D. V. (2007). Environment - Carbon mitigation by biofuels or by saving and restoring forests. *Science* 317 (5840), 902–902.

Robin, E. (2005).http://www.arqsustentable.net/actualidad_video_soja.html

Rodriguez, A.M. and Jacobo, E.J. (2010). Glyphosate effects on floristic composition and species diversity in the Flooding Pampa grassland (Argentina). *Agriculture, Ecosystems and Environment.* doi:10.1016/j.agee.2010.05.003.

Ryan, P. and Openshaw, K. (1991). Assessment of biomass energy resources: a discussion on its need and methodology. Report Nº 11338. Industry and Energy Department Working paper. *Energy Series Paper* No. 48.

Salomon, O.D., Orellano, P.W., Quintana, M.G., Perez, S., Sosa, E.S., Acardi, S. and Lamfri, M. (2006). Transmisión de la Leishmaniasis Tegumentaria en la Argentina. *Medicina* 66, 211-219. Buenos Aires.

Sanhueza, E. (2009). Agroetanol ¿Un combustible ambientalmente amigable? *Interciencia* 34 (2), 106-112.

Sarandón, S.J and Iermanó, M.J. (2005a). Sustainability of the production of biodiesel as an alternative fuel In Argentine Republic. Anales (CD-rom) III Congreso Brasileiro de Agroecologia, III Seminario Estadual de Agroecología, Florianópolis, 17 al 20 de Octubre de 2005, Florianópolis, SC, Brasil. 25, 4pp.

Sarandón, S.J. and Iermanó, M.J. (2005b). Energetyc efficiency of biodiesel production from different oil crops: an agroecological analysis. Anales (CD-rom) III Congreso Brasileiro de Agroecología, III Seminario Estadual de Agroecología, Florianópolis, 17 al 20 de Octubre de 2005, Florianópolis, SC, Brasil. 601, 4pp.

Sasaki, N., Asner, G.P., Knorr, W., Durst, P.B., Priyadi, H. and Putz, F.E. (2010). Approaches to Classifying and Restoring Degraded Tropical Forests for the Anticipated REDD+ Climate Change Mitigation Mechanism. iForest - Biogeosciences and Forestry.

SAyDS (Secretaría de Ambiente y Desarrollo Sustentable). (2005). Primer Inventario Nacional de Bosques Nativos. Proyecto Bosques Nativos y Áreas Protegidas. Préstamo BIRF 4085-AR; 2005. Informe Nacional. Argentina.

SAyDS (Secretaría de Ambiente y Desarrollo Sustentable de la Nación). (2008). El avance de la frontera agropecuaria y sus consecuencias. Buenos Aires, Argentina.

Schnoor, J.L., Doering, O.C., Entekhabi, D., Hiler, E.A., Hullar, T.L. and Tilman, D. (2008). Water implications of biofuels production in the United States. National Academy of Sciences, Washington DC, USA. <http://www.nap.edu/catalog/12039.html>.

Schubert, R. and Blasch, J. (2009). Sustainability standards for bioenergy - a means to reduce climate change risks. Earth and Environmental Science 6 (2009) 182003 doi:10.1088/1755-1307/6/8/182003.

Searchinger, T., Heimlich, R., Houghton, A., Dong, F., Elobeid, A., Fabiosa, J., Tokgoz, S., Hayes, D. and Yu, T. (2008). Use of U.S. Croplands for biofuels increases greenhouse gases through emissions from land-use change. *Science* 319, 1238.

SEN (Secretaría de Energía de la Nación). (2006). Balance Energético Argentino. Versión preliminar. Buenos Aires, Argentina.

Seghezzo, L. (2008). Desarrollo a Secas, Desarrollo Sustentable y Sustentabilidad. Curso de Postgrado. Facultad de Ciencias Naturales. Universidad Nacional de Salta. Argentina.

Seijo, A. (2008). Boletín de temas de salud de la Asociación de Médicos Municipales de la Ciudad de Buenos Aires. Suplemento del Diario del Mundo Hospitalario, Año 15 (138). http://www.medicos-municipales.org.ar/bts0708.htm.

Semino, S. (2008). Can certification stop high soy pesticide use? *Pesticide News* 82, 9–11.

Semino, S., Paul, H., Tomei, J., Joensen, L., Monti, M. and Jelsøe, E. (2009). Soybean biomass produced In Argentina: myths and realities. Beyond Kyoto: addressing the challenges of climate change. *Earth and Environmental Science* 8, 1-13.

Sheehan, J.J. (2009). Biofuels and the conundrum of sustainability. *Current Opinion in Biotechnology* 20, 318–324.

Shi, X., Elmore, A., Li, X., Gorence, N.J., Jine, H., Zhang, X. and Wang, F. (2008). Using spatial information technologies to select sites for biomass power plants: A case study in Guangdong Province, China. *Biomass and Bioenergy* 32, 35– 43.

Smeets, E.M.W., Faaij, A.P.C., Lewandowski, I.M. and Turkenburg, W.C. (2007). A bottom-up assessment and review of global bio-energy potentials to 2050. *Progress in Energy and Combustion Science* 33, 56–106.

Solomon, D.S. and Hughey, K.F.D. (2007). A proposed Multi Criteria Analysis decision support tool for international environmental policy issues: a pilot application to emissions control in the international aviation sector. *Environmental Science and Policy* 10, 645-653.

Takada, M., Morris, E. and Rajan, S.C. (2000). Sustainable Energy Strategies. Materials for decision makers. United Nations Development Programme. Initiative for Sustainable Energy (UNISE).

Tijmes, P. and Luijf, R. (1995). The Sustainability of our Common. Future: An Inquiry into the Foundations of an Ideology. *Technology in Society* 17 (3), 327-336.

Tomei, J. and Upham, P. (2009) .Argentinean soy-based biodiesel: an introduction to production and impacts. Tyndall Working Paper no.133. /http://tyndall.ac.uk/ publications/working_papers/twp133.pdfS.

Tsoutsos, T., Drandaki, M., Frantzeskaki, N., Iosifidis, E. and Kiosses, I. (2009). Sustainable energy planning by using multi-criteria analysis application in the island of Crete. *Energy Policy* 37, 1587–1600.

Tuesca, D., Niesemsohn, L. andPapa, J. (2007). Para estar alerta: el sorgo de alepo resistente al glifosato. INTA/EEA: Oliveros, Argentina: http://www.inta.gov.ar/oliveros/info /documentos/soja/soja_malezas1.pdfS

UNEP (United Nations Environment Programme). (2009). Global Trends in Sustainable Energy. Investment 2009. Analysis of Trends and Issues in the Financing of Renewable Energy and Energy Efficiency. UNEP's Division of Technology, Industry and Economic (DTIE) under its Sustainable Energy Finance Initiative and was produced in collaboration with New Energy Finance Limited. 64 p.

USDA (United States Department of Agriculture). (2008). Argentina soybeans and products supply and distribution local marketing years.USDA Foreign Agricultural Service, http://www.fas.usda.gov/psdonline/psdreport.aspx? hidReportRetrievalName=BVSandhidReportRetrievalID=722andhidReportRetrievalTemplateID=13S.

Van Dam, J., Faaij, A.P.C., Hilbert, J., Petruzzi, H. and Turkenburg, W.C. (2009). Large-scale bioenergy production from soybeans and switchgrass in Argentina. Part B. Environmental and socio-economic impacts on a regional level. Renewable and Sustainable *Energy Reviews* 13, 1679–1709.

Von Braun, J. (2008). Rising food prices: What should be done? International Food Policy Research Institute, Washington DC, USA. <http://www.ifpri.org/pubs/bp/bp001.asp>.

WCED (World Commission on Environment and Development). (1987). Our Common Future. Oxford: Oxford University Press.

Zak, M.R., Cabido, M., Caceres, D. and Diaz, S. (2008). What drives accelerated land cover change in central Argentina? Synergistic consequences of climatic, socio- economic and technological factors. *Environmental Management* 42,181–189.

Zidansek, A., Blinca, R., Jeglic, A., Kabashi, S., Bekteshi, S. and Slaus, I. (2009). Climate changes, biofuels and the sustainable future. *International journal of hydrogen energy* 34, 6980–6983.

In: Bioenergy Systems, Biological Sources … ISBN: 978-1-62417-330-1
Editor: Michel C. Allard

Chapter 2

Soil Loss – A Forgotten Aspect of Biofuel Production

Michael W. H. Evangelou* and Rainer Schulin
Institute of Terrestrial Ecosystems, ETH Zürich,
Zürich, Switzerland

Abstract

The increasing demand for energy and the decreasing reserves of fossil fuels have turned the focus of governments to the production of energy from biomass (biofuels, biogas). Since biomass is currently the only renewable source of fixed carbon it is an ideal source for the production of transport fuels. However, because of the high production targets set by various governments, much fertile agricultural land is diverted from food to bioenergy production. Moreover, because of agricultural practices, sol is lost or degraded. Soil is one of the most important but equally neglected resources. It not only produces food/forage/bioenergy but also sequesters greenhouse gases, filters rain water before reaching the groundwater and supports biodiversity. It is important to select suitable bioenergy plants, such as perennial grasses and short rotation wood coppice instead of row crops such as corn. Furthermore, attention should be given to no tillage practice with an incorporation of crops residues into the soils, thereby increasing the sustainability of biofuel production. In addition, soils which are already degraded could be used as an alternative to fertile land. Attention should be given to not only conserving land, but to improving and perhaps even restoring contaminated and degraded soils in the long term.

1. Introduction

"Green energy" in the form of biomass has been greeted as a major renewable energy source in supplementing declining fossil fuel resources. It is currently the only renewable

* Corresponding author: Institute of Terrestrial Ecosystems, ETH Zürich, Universitätstrasse 16, CH-8092 Zürich, Switzerland. Tel: +41-44-6327402, e-mail: michael.evangelou@env.ethz.ch

source of organic carbon, thus making it the only long term source for the production of transport fuels (Bridgwater 2006). Using biomass for fuel production is seen as an attractive option for three main reasons. Firstly, it is a renewable resource und thus sustainable. Secondly, because there is no net release of carbon dioxide and sulfur contents are low, it is environmentally attractive. Thirdly, given the rise in fossil fuel prices, it is also economically advantageous (Demirbas 2007).

As many countries strive to become less reliant on foreign oil, they aim to increase renewable energies production, including biofuels. In the last few years several countries have published directives and mandates stating their goals for biofuel production by 2020-2022. The Renewable Energy Directive (Parliament and Council 2009) (2009/28) of the EU specifies that by 2020 each member state has to achieve at least 10% of their transport fuel consumption from renewable sources – 7% of that will probably come from biofuels. The amount of biofuels targeted for 2020 would be approximately 6.5 billion gallons (1.7 billion liters). In 2007, the U.S. Congress passed the Energy Independence and Security Act (Congress 2007) (EISA; U.S. Congress 2007), which sets targets for the proportion of transportation fuels that are to be renewable including conventional biofuels (e.g., corn-based ethanol) and "additional renewable fuels" (e.g., advanced biofuels, biomass-based diesel, and cellulosic). The total amount of renewable fuels targeted by 2022 would be 36 billion gallons (9.5 billion liters) or approximately 13% by volume of all transportation fuels. China aims at accounting for 15 percent of energy consumption through non-fossil fuels by the end of 2020. Compared with other renewable energy sources such as hydropower, solar energy and wind power, biofuel is expected to play a minor part in China's diversified energy policy because of a lack of sustainable supply of feed stocks and technology breakthrough (Scott and Junyang 2011). Nevertheless, the production of biofuels, such as ethanol, is subsidized by the Chinese government and the biofuel production could be expected to reach approximately 3 billion gallons (0.8 billion liters). Further 49 countries including India, Brazil, Uruguay and the Philippines have passed similar mandates resulting in a global biofuel demand of approximately 60 billion gallons (15.8 billion liters).

The subsidies of biofuel crops by many countries, including the United States, the European Union, Brazil and China has increased the public profile of biofuels in recent years. To reach the biofuel demand of 60 billion gallons (15.8 billion liters) vast swathes of land from would have to be converted from food to energy crop production. Similarly forests and grasslands would need to be cleared to enable the production of energy crops. The decrease in food production arising from the popularity of energy crops, combined with a rising demand for food and a lower availability of arable land (Figure 1) in an increasing world population could increase the cost of food commodities which are critical to the diets of food-insecure people worldwide.

The OECD/FAO (2011) estimated that by 2020, 12% of the global coarse grain production as well as 33% of the sugar production will be used to produce ethanol. Additionally, 16% of the global production of vegetable oil will be used to produce biodiesel. The conversion of forests to arable land for bioenergy crops could result in CO_2 emissions from terrestrial carbon pools that are much greater than any greenhouse gas benefits provided by biofuels.

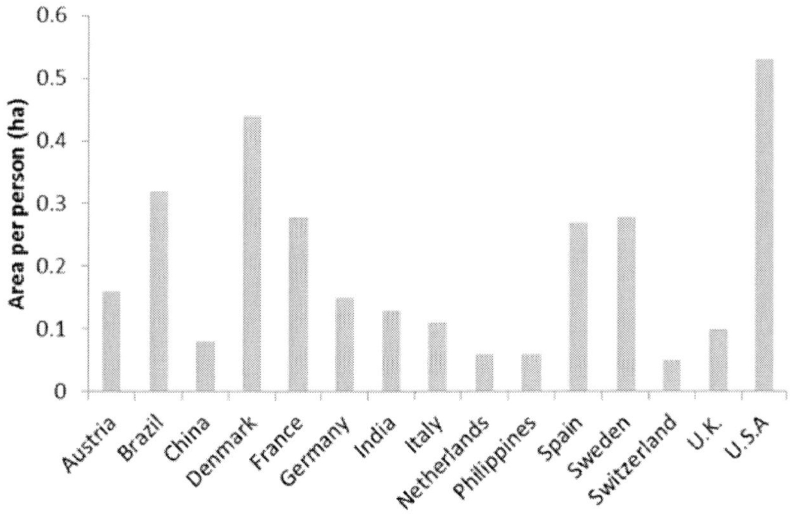

Figure 1. Estimation of arable land per person (TWB 2012) for major biofuel producing countries.

Converting rainforests, peatlands, savannas, or grasslands to produce food crop–based biofuels in Brazil, Southeast Asia, and the United States creates a "biofuel carbon debt" by releasing 17 to 420 times more CO_2 than the annual greenhouse gas reductions which these biofuels would provide by displacing fossil fuels (Fargione et al. 2008; Righelato and Spracklen 2007).

Owing to the above mentioned issues of food security and climate change, and the subsequent loss of biodiversity, it has been argued that biomass production for biofuels is currently unsustainable (Friedemann 2007). As a result some studies have suggested using marginal, idle and degraded lands (Campbell et al. 2008; Milbrandt and Overend 2009), as well as contaminated land (Evangelou et al. 2012a; Robinson et al. 2009) in the form of phytomanagement. Frequently, investigations, by focusing on the efficient use of land for food and the dangers of climate change, have neglected the importance of soil itself as a resource.

Soil is a valuable resource and its importance does not lie exclusively in its role as the indispensable production factor for agriculture and forestry, but in a variety of functions, such as the maintenance of biological diversity (e.g. as the habitat for organisms), protection against natural hazards (e.g. by providing stability against erosion, particularly in alpine regions), hydrological balance functions (e.g. water storage), acting as a means of filtering out pollutants and as a long-term carbon sink. Despite being crucial, these functions are frequently undervalued.

About 2 000 million ha of soil, equivalent to 15 per cent of the Earth's land area (an area larger than the United States and Mexico combined) have been degraded because of human activities. The main types of soil degradation are water erosion (56 per cent), wind erosion (28 per cent), chemical degradation (12 per cent) and physical degradation (4 per cent).

The causes of soil degradation include overgrazing (35 per cent), deforestation (30 per cent), agricultural activities (27 per cent), overexploitation of vegetation (7 per cent) and industrial activities (1 per cent) (UNEP 2002). The threat to sustainable development posed by land degradation has been recognized for decades, and was discussed at the 1992 Earth Summit and the 2002 World Summit on Sustainable Development.

There have been ongoing developments to reduce soil degradation, by givinggreater attention to biological methods of conservation, and the integration of water conservation with soil protection. Attempts have been made to improve the management of soil plant-water relationships, including reduced disturbance by tillage. Nevertheless, there is no clear indication that the rate of land degradation has decreased (UNEP 2002).

With increased biofuel production land degradation could become more severe in the near future. In Indonesia for instance, two-thirds of oil palm expansion has occurred by converting large rainforest areas. Recently, in the United States 1.3 million ha of set-aside lands in the Conservation Reserve Program designed to help check surpluses, maintain price levels and promote an ecological balance were called back into production (UNEP 2012). Such shifts illustrate the complexities involved. The sustainable use of soil resources has to become the main focus when producing biofuels, so that in the long-term the environmental quality of soil would not just be conserved but also improved or even restored.

2. Impact of Biofuel Production on the Environmental Quality of Soil

The effect which the production of biomass for biofuels has on the environmental quality of soil depends on the choice of plants (e.g. crop, woody species, perennial grasses), the soil type (e.g. sandy, clay), the management (e.g. tillage, fertilizer application), the climate (e.g. dry, humid) and the geomorphology (e.g. slope). The main impacts are on soil erosion, soil organic carbon (SOC), soil structure, nutrients and biological properties.

2.1. Soil Erosion

Soil erosion is a natural process, occurring over geological time, and indeed it is a process that is essential for soil formation in the first place. Owing to anthropogenic activity, soil erosion has accelerated, thus constituting one of the main drivers for soil degradation. Soil erosion has contributed to the loss of 430 million ha of cropland worldwide, 30% of the world cropland base. Soil loss negatively affects characteristics associated with crop productivity including water holding capacity, soil nutrients, soil density, soil organic matter (SOM) and others (Kort et al. 1998). Soil is being lost worldwide from land areas 10 to 40 times faster than the rate of soil renewal. This endangers both future food supplies and the environment. One of the main biofuel producers, the United States is losing annually 3 billion tons, a rate of 10 times faster than the natural replenishment rate. In the cases of China and India the amounts are 5.5 billion tons per year 6.6 billion tons per year, which equals to a rate of 30 and 40 times faster than the natural replenishment rate, respectively.(Pimentel 2006). The economic impact of soil erosion costs the United States alone approximately \$37.6 billion each year in productivity losses. The damage worldwide is estimated to be \$400 billion per year. With the increasing production of biomass crops for biofuels and the conversion of forests and grassland for these purposes, soil erosion can be expected to increase unless appropriate measures are undertaken.

2.1.1. Choice of Plants

Several plant species are considered for the production of biomass for bioenergy, including crops such as corn (*Zea mays*) or wheat (*Triticum* spp.), trees such as poplar (P*opulus* spp.) and willow (*Salix* spp.) as well as perennial grass species such as switchgrass (*Panicum virgatum*). These species can either reduce or increase soil erosion depending on their root structure, the soil cover they offer, their effect on SOC content and agricultural management. Sullivan (2004) showed that traditional annual crops such as corn and soybean (*Glycine max*) cause 50 times more soil erosion than sod crops such as hay because the soil between the rows can wash or blow away. Tree species, such as willow or poplar have been planted for many years to reduce water erosion by creating a living buffer area adjacent to water courses. By such means sediment is caught from eroding land, water infiltration is increased, and stream banks are stabilized by the tree roots (Kort et al. 1998). Their effect on reducing soil erosion could be seen in a study by Pimentel and Krummel (1987) in which the authors determined the average erosion rate under short rotation woody crops (SRWC) and corn. Under SRWC the average erosion rate was 2 Mg ha^{-1} yr^{-1} on a 5% slope, whereas corn production on a 4% slope resulted in a soil loss of 21.8 Mg ha^{-1} yr^{-1}. Perennial grasses (PG) have also been shown to be quite effective in controlling erosion (Kemper et al. 1992). Harvesting does not necessarily negate their erosion control benefits because they maintain considerable biomass below the typical cutting or grazing heights. Furthermore, perennial grasses have a network of fibrous roots in surface layers of soil which aid in erosion control. The species choice, however, is of critical importance. Switchgrass for example has a seven-fold greater root mass in the surface 0-15 cm than orchardgrass (*Dactylis spp.)*. It also displays roots as deep as 60 cm whereas the roots of orchardgrass do not reach below a depth of ca. 30 cm (Kort et al. 1998). Both PG and SRWC can also control wind erosion. Perennial grasses such as switchgrass resist erosive forces of wind because they grow tall and have stiff stems and deep rooting systems. In semiarid regions where wind erosion is of greater magnitude than water erosion, the establishment of PG can be particularly effective for controlling wind erosion near the soil surface. By contrast trees, can reduce wind velocity at increased heights above the soil surface (Blanco-Canqui and Lal 2008).

Although SRWC have been shown to display a lower risk to soil erosion than crops, herbaceous undergrowth is still needed. Without herbaceous cover the hazards remain high especially when rainfall downfalls are high (Kort et al. 1998). Depending on the tree species and the groundwater availability, the water table of a SRWC plantation can be lower than in adjacent land. This can have a positive effect, as was the case in Australia, where Eucalyptus was used to lower the yearly minimum groundwater table in order to reduce salinization of the soil by excessively shallow groundwater (Bari and Schofield 1992). The depletion of groundwater may, however, also have negative consequences, especially after biomass harvesting: it could make the soil more susceptible to wind erosion. The choice of plants can also influence the agricultural practice, which subsequently can affect soil erosion (see 2.1.2). Perennial energy crops require less tillage than annual food crops, since they do not need to be re-established. They are thus less affected by certain agronomic limitations which can complicate tillage operations (Nijsen et al. 2012).

2.1.2. Agricultural Management

As an agricultural practice, one can distinguish, two main factors that affect soil erosion: the utilization of crop residues and the application of tillage or no tillage practice. Crop residues, mainly corn stover, are considered to be the prime lignocellulosic feedstock for producing ethanol. The goal of energy industries is to use crop residues as cellulosic ethanol feedstock in the short term while other biomass sources from dedicated energy crops are being developed in the long term. Crop residues left on the soil surface are vital for soil protection from wind erosion, particularly in semiarid regions. If residues are completely removed, no-tillage soils can be as erodible as or more than plowed soils during drought periods, because of absence of transient rough surface created by tillage (Blanco-Canqui 2010). As a compromise, Blanco-Canqui and Lal (2009b) propose that a partial removal of 25% of stover may be the threshold removal rate from erosion-prone soils. However, they also state that logistics of harvesting stover by 25% need a careful and objective consideration. Additionally, they estimate that a removal of only 25% of stover may neither be economically feasible nor sufficient to produce high volumes of ethanol needed to ease dependence on fossil fuels.

No-tillage is a conservation tillage practice that, in addition to leaving considerable residue on the soil surface, leaves the soil undisturbed from harvest to planting. This promotes well-formed soil macroaggregates which result in increased infiltration, less runoff, and less soil movement (Williams et al. 2009). Although, there are several tillage techniques which have various effects on soil erosion, in general one can say that tillage increases the risk of soil erosion because of the destruction of soil structure, of vertical pore system created by roots, or earthworm and other soil animals, the acceleration of organic matter mineralization (depletion) and the reduction of aggregate stability.

There are several studies that show that no-till cropping systems are more effective in reducing runoff and soil erosion than conventional tillage cropping systems (Williams et al. 2009).

2.2. Soil Organic Carbon

Soil organic carbon (SOC), the major component of SOM, is extremely important in all soil processes. It has the capacity to affect plant growth as both a source of energy and a trigger for nutrient availability through mineralization. It is the main source of energy and nutrients for soil microorganisms. Components of the SOC, such as glomalin also stabilize soil structure by gluing together aggregates, thus making soil resistant to erosion, while still being porous enough to allow air, water and plant roots to move through the soil. Thus, a poor SOC content reduces microbial biomass, activity and nutrient mineralization. It can reduce drainage, aggregate stability, infiltration and aeration in soils. Conversion of native ecosystems to crops causes substantial losses of SOC (averaging about 30%), which is most rapid immediately following land conversion probably stabilizing within the first several decades of cultivation. Overall, approximately 55–78 GtC have been released from soil as a result of land conversion during the postindustrial era. Conversely, SOC accumulates upon conversion of cultivated land to secondary forest, grassland, or pasture. Because grasslands tend to have high SOC, conversion of native ecosystems to pasture generally does not release SOC, and sometimes results in modest SOC sequestration (Anderson-Teixeira et al. 2009).

Amongst the several factors, which can influence the SOC fluxes are climatic conditions, soil type and SOC content. Those, though, which can be most easily controlled to avoid a loss of SOC are the plant species utilized for biofuel production and the agricultural management implemented.

2.2.1. Choice of Plants

The choice of the plant species for biofuel production can have a great influence on the SOC content, because in time the partitioning of C within the plant affects how much C becomes harvestable biomass and how much enters the soil. As roots and rhizomes are a primary source of C to the soil in biofuel crops, belowground allocation of C stands to play a critical role in driving changes in SOC. Allocation of C to root biomass is dramatically lower in traditional annual crops than in temperate grasslands, with globally averaged root : shoot ratios of 0.1 and 3.7, respectively. The following biofuel crops match this pattern; observed root : shoot ratios are 0.3 for fertilized corn, 0.15 for sugarcane in nonfield conditions, 1 for Miscanthus, 1.8–6.1 for switchgrass and 2–8 for North American prairies. Thus, the temperate PG introduce far more C to the soil than corn, potentially explaining the difference between corn and these PG (Anderson-Teixeira et al. 2009). This allocation of C to the root can subsequently affect the SOC content. Elustondo et al. (1990) reported a 25-29% increase in the SOC content of the surface soil (0 - 7.5 cm) of a continuous meadow relative to 5 years of continuous corn. McLaughlin and Walsh (1998) reported that carbon sequestration rates under switchgrass may exceed those of annual crops by as much as 20-30 times, owing in part to carbon storage in the soil. In contrast to corn, cultivation of temperate-zone PG – Miscanthus, switchgrass, or native mixes – can increase SOC by an average of approximately 0.1–1 Mg ha^{-1} yr^{-1} (Anderson-Teixeira et al. 2009). Short rotation wood coppice, such as willow has shown to be even more effective in storing SOC than switchgrass. In a study by Zan et al. (2001), on relatively fertile soils in Canada, willow trees stored more SOC than either corn or switchgrass after 4 yr. of establishment, although switchgrass and willow had similarly high root biomass C. The higher SOC accumulation could be due to the fact that native PG have been shown to undergo complete root turnover only once every 4 years compared with at least once per year for fine roots of deciduous tree species, which release substantial amounts of C upon their decay (Zan et al. 2001). Besides the amount of SOC stored, the choice of plant species also plays a role where the SOC is stored, as the SOC fraction stored in deeper (>30 cm) soil profiles has longer residence times and slower turnover due to reduced microbial processes and fluctuations in soil water content and temperature, thus affecting long-term SOC sequestration.

2.2.2. Agricultural Management

Perennial grasses and SRWC as bioenergy crops generally increase the SOC pools, but, the magnitude of increases depends on harvest cycles, tillage practice, crop residues management, soil type, and application of N fertilizers and manure. Reduction in frequency of harvest increases SOC sequestration because of greater C biomass return (Blanco-Canqui 2010). Tillage releases SOC and is probably largely responsible for SOC losses during land conversion. Conversely, reduced tillage results in soil C sequestration and SOC accumulation under PG grown on former cropland may be partially attributable to the cessation of tillage (Anderson-Teixeira et al. 2009). Crop residue management can also play an important role in SOC pool changes. Crop residue removal can cause negative SOC pool changes, i.e. loss of

SOC, while crop residue removal or return can cause positive SOC pool changes i.e. gain of SOC. The importance of crop residue management for SOC was made clear by Wilhelm (2007) who found that the amounts of corn stover needed to maintain SOC, and thus productivity, exceed by far the amounts needed to control water and wind erosion. Although the above mentioned pool changes are more or less true for most soils, the actual amounts gained and lost depend on several factors, such as SOC concentration, slope and soil texture. Blanco-Canqui (2010) reported that changes in SOC concentration due to crop residue removal can be slower in nearly level clay soils than in sloping and erosion-prone soils. On silt loam with 10% slope, silt loam with 2% slope, and clay loam with <1% slope, Blanco-Canqui and Lal (2009a) reported that stover removal at rates as low as 50% reduced SOC concentration by nearly 5.5 Mg ha^{-1} in silt loams in the 0- to 10 cm soil depth after 4 yr. of stover removal, showing that SOC pool was reduced by 1.4 Mg ha^{-1} yr^{-1}. On the clay loam, stover removal at 50% did not significantly reduce SOC concentration, but removal at 75% reduced SOC concentration by 6.5 Mg ha^{-1} for the same soil depth, indicating a decrease of SOC pool by 1.6 Mg ha^{-1} yr^{-1}. Complete stover removal reduced SOC concentration by 3 Mg ha^{-1} yr^{-1} in the sloping silt loam, by 2.2 Mg ha^{-1} yr^{-1} in the nearly level silt loam, and by 1.6 Mg ha^{-1} yr^{-1} in the clay loam. In some soils, SOC pools under stover removal or return may be unaffected if the soils have reached a steady-state SOC level. Hooker et al. (2005) reported that after 29 yr. of no tillage management, no differences in SOC pool between soils with and without stover mulch could be detected.

2.3. Soil Structure

Soil structure is a complex condition that is related to many agronomic and environmental processes. It has a great influence on biogeochemical processes. Soil microbial processes are directly and indirectly influenced by soil structure. The presence of small pores reduces accessibility of organic materials to decomposers, causing the physical protection of C and a reduction in N mineralization. Soil structure also exerts a strong influence upon soil hydraulic and solute transport processes. Additionally, the distribution of soil structural units controls the availability of oxygen, and the resistance to penetration by roots (Díaz-Zorita et al. 2002). Soil structure can be influenced positively by the choice of suitable cover crops, an agricultural management which aims at avoiding compaction, and which optimizes irrigation and incorporates a suitable residue and tillage management.

2.3.1. Choice of Plants

Growing PG as compared to row crops generally improves soil structural properties. It reduces bulk density, increases soil porosity, improves soil aggregation (Blanco-Canqui 2010). Bharati et al. (2002) observed that soil bulk density under switchgrass grown in a riparian buffer was 83% of that under corn, while Rachman et al. (2004) observed that after 10 years of switchgrass establishment the bulk density was reduced, as compared to a no-tillage corn-soybean rotation. In the same study Rachman et al. (2004) also observed that soil macroporosity under switchgrass was about two times greater than under corn-soybean rotation. Furthermore, PG build an abundant and dense network of roots, which forms a skeleton enmeshing soil particles and forming stable aggregates. The growth of SRC trees has also positive effects on soil structural properties compared to row crops. In a study by Kahle et al. (2005) it was reported that the afforestation of cropland soils with willow and poplar decreased soil bulk density, increased soil porosity (especially the proportion of medium pores) and improved soil water retention after 6 yr. of conversion. The effect that trees and PG have on soil structure influences also soil hydraulic properties, which include infiltration, hydraulic conductivity and water storage. In comparison to row crops, PG display greater macroporosity, more fibrous and extensive roots, more abundant and interconnected macropore networks (e.g., root channels and earthworm burrows), thus, improving infiltration, hydraulic conductivity and water storage. With the gradual development of biopores, root growth, and macroaggregates following the establishment of PG the rate at which water moves in lower soil profile depths improves (Blanco-Canqui 2010). Bharati et al. (2002) observed that water infiltration was five times greater under multispecies riparian buffer, consisting of silver maple, grass filter, and switchgrass than under the cultivated field (corn) and pasture after 6 yr. of management. The tree species can be a very important aspect concerning water infiltration. For example the water infiltration rate under a 3-yr-old Eucalyptus plantation was as high as that under a 27-yr-old pine plantation (Kort et al. 1998).

2.3.2. Agricultural Management

Tillage induces significant temporal and spatial changes in soil structure and thus the growing conditions for plants. The repeated tillage of agricultural soils disrupts soil macroaggregates (>2 mm) and creates less stable soil aggregates. It may produce more homogeneous soil in the tilled layer, but it also induces heterogeneity in the form of compaction under the tilled zone and under the wheels of the tractor pulling the tillage implement (Munkholm et al. 2008). Although conventional tillage often degrades soil structure the incorporation of crop residues into farming systems could improve soil structure, reduce bulk density and improve water infiltration (Kort et al. 1998). Under SRWC, the removal or return of litter to the soil can also have an effect on soil structure. The presence of tree litter on the soil surface at various stages of decomposition protects soil against raindrop impact and minimizes crusting or surface sealing. Decomposed and partly decomposed organic materials reduce risks of soil compaction by reducing internal friction and consolidation of soil aggregates (Blanco-Canqui 2010). Indiscriminate removal of residues can increase the soil's susceptibility to compaction, crusting, and surface sealing. It can reduce soil aggregation, macroporosity, aeration, and water infiltration (Wilhelm et al. 2004). Blanco-Canqui and Lal (2009b) observed that stover removal at rates as low as 25% reduced the stability of aggregates in nearly level silt loam and clayey soils. Stover removal can also cause an increase in soil compaction and reduce the plant available water content of the soil. The effect that the stover removal can have on the plant available water content of the soil depends on the soil texture. The plant available water decreased linearly with an increase in the rate of stover removal (25%-100%) for silt loam and quadratically for clay soils (Blanco-Canqui and Lal 2007).

2.4. Soil Biology

Soil microorganisms such as bacteria and fungi and as soil fauna such as earthworms and springtails have very important soil processing roles. They play a vital part in the decomposition of organic residues, the subsequent creation of humus and the release of nutrients from the decaying organic residues. They are important for the aeration and drainage of the soils through the creation of tunnels, as well as the incorporation of organic matter in deeper soil layers. They are also of direct importance for plant growth as they can form partnerships with plants, such as the nitrogen-fixing bacteria. The diversity and the biomass of soil organisms depends on the soil management, the plants, the soil itself and climatic conditions. Doran (1987) found that microbial populations and activities were regulated more by crop type and rotation than by soil physical properties.

Hence with a suitable choice of plants and soil management it would be possible to positively influence soil organism biomass and diversity.

2.4.1. Choice of Plants and Agricultural Management

Increased food supply (e.g., partly decomposed organic materials), adequate soil water content, and reduced fluctuations in soil temperature under growing PG and SRWC promote proliferation of soil macro- and microorganisms. In loamy soils with exposed subsoil, Al-Kaisi and Grote (2007) found that switchgrass had 200% more microbial biomass than either corn or soybean. In their investigations Ma et al. (2000) found an increase of 168% of microbial biomass C content 2 yr. after switchgrass establishment on previously cropped sandy loam soil.

Bharati et al. (2002) observed that earthworms under trees and grass buffers were often more numerous than under croplands because of the abundant food source and cover. This observation was confirmed in a study by Katsvairo (2007) which showed that in a sandy loam, earthworm numbers (6.3 m^{-2}) were greater under switchgrass than in the adjacent (0 m^{-2}) peanut (Arachis hypogaea L.)-cotton cropping system.

In the above mentioned studies the influence of the agricultural practice cannot be neglected. A change in plant species could subsequently lead to a change in tillage practice.

2.4.2. Agricultural Management

Cropping systems can themselves influence the biomass of macro- and microorganisms. Tillage may have detrimental effects on soil organisms, and by itself, it is associated with reduced fungal biomass which in turn can affect fungal-feeding nematodes (Treonis et al. 2010).

Arbuscular mycorrhizal (AM) fungi, in particularl may be very sensitive to tillage (Simmons and Coleman 2008). Speeding (2004), however showed that the effect of tillage is not so pronounced as the effect of incorporating plant residues, which increased soil microbial biomass C and N. Acosta-Martinez et al. (2007), support these findings. They reported higher soil enzyme activities, microbial biomass and fungal populations with a combination of no-till and increased cropping intensity, potentially as a result of greater residue quantity.

Crop residues, and other organic amendments such as manure or cover crops, are often incorporated into soils though tillage. Thus, their effect, of increasing the abundance of various components of the soil food web, including the soil microbial community, protozoa, and microbivorous nematodes should expand into deeper soil layers.

Treonis et al. (2010) reported that in control soils, tillage reduced the relative abundance of fungal-feeding nematodes and increased the density of bacterial-feeding nematodes. When combined with organic amendment however, tillage was associated with increases in fungal-feeding nematodes and fungal biomass, as well as with declines in plant-parasitic nematodes.

3. Ecological Aspects

In several cases attempts at improving soil quality by growing biomass plants, such as the Eucalyptus or the Miscanthus, have produced problems: The proposed plant species have become invasive species/neophytes in some parts of the earth. Bioinvasion costs the United States (Bright 1998) as well as other countries (Hulme et al. 2009) billions of dollars per year. For instance, Johnson grass, introduced as a forage grass, is now an invasive weed in many states.

Another fast-growing perennial grass, Miscanthus, already used for biofuel production, has been described as ''Johnson grass on steroids'' (Raghu et al. 2006). Growing non-invasive species might initially be less economically attractive. However, in the long term they whould be more advantageous as no clearing costs would be involved, and public opinion would be less hostile.

When designing a plantation and selecting plant species it is imperative to consider surrounding ecosystems. It is essential that the establishment of a biomass for biofuel plantation does not have detrimental effects.

In a study by, Cornish (1989), it was discovered that, although Radiata pine (*Pinus radiata*) planted in Australia reduced soil erosion and increased slope stability, it also reduced stream flow so effectively that naturally perennial streams were turned into ephemeral streams. There was thus a possibility, that habitats of several fauna species requiring perennial stream flow would be endangered.

4. Alternative Lands for Biofuel Production

4.1. Degraded Soils as a Resource

A sustainable use of soil resources is possible provided that both the choice of plants and agricultural management are suitable conserving, environmental quality of fertile soils. The possibility of suing degraded soils, thereby reserving fertile soils for food production, should not be forgotten. According to a study by Nijsen et al. (2012), the total global area of degraded land equals 1836 Mha. The degraded area is spread across all regions in the world. As much as 15% is found in China. Other regions that have significant areas of degraded land are the Middle East (9%), India, Brazil and the region Rest of South America. The last three regions each account for 6% of the global area of degraded land. However, most (46%) of the degraded land is moderately degraded, whilst 37% is classified as lightly degraded and 15% and 0.5% as strongly and extremely degraded, respectively. Not all of these soils should be considered potential areas for biofuel production because slightly and moderately affected soils might still be suitable for the production of food, especially in land scarce regions (Nijsen et al. 2012). The total area of degraded land that is potentially available for biofuel production has been estimated by Nijsen et al. (2012) as 247 million ha. Campbell et al. (2008) put the figure significantly higher at 385-472 million ha, and Tilman et al. (2006) estimated it as 500 million ha.

Expectations for the amount of biofuel production are perhaps too much, as the yields will vary according to the fertility of the soils. Estimates vary greatly. Nijsen et al. (2012)

expect a bioenergy potential of 25-32 EJ, Campbell et al. (2008) of 32-41 EJ, van Vuuren et al. (2009) of 43 EJ and Tilman et al. (2006) of 45 EJ. According to the calculations by Campbell et al. (2008) at the national scale, the bioenergy potential is less than 10% of primary energy demand for most countries in North America, Europe, and Asia. However, it also represents many times the current energy demand in some African nations where grasslands are relatively productive and current fossil fuel demand is low. Thus, for the main biofuel producers, the use of only the degraded land for biofuel production will not be sufficient to fulfill the bioenergy demand. However, if the aim is not just biofuel production but also the improvement or even the restoration of degraded soil, then with suitable bioenergy plants and agricultural management one could have a win-win situation (Lal and Pimentel 2007). Furthermore, with the improvement of soil quality, the yields would increase thus covering a higher fraction of the primary energy demand.

The majority of the soils are degraded by wind and water erosion (UNEP 2002). As a result, PG and SRWC with green cover should be preferred to the erosion prone row crops such as corn. Furthermore, a no tillage practice, as well as the incorporation of crop residues should be preferred (Table 1).

The no tillage practice would subsequently favor perennial energy crops as they require less tillage than annual food crops, because perennial crops do not need to be re-established. The content of SOM is vital for plant yield. Lal (2006) showed that for every additional t C in SOM yields increased greatly: 20–70 kg ha-1 yr-1, for wheat, 10–50 kg ha1 yr-1 for rice, and 30–300 kg ha-1 yr-1 for maize. Consequently, plants, such as PG should be preferred, as they can increase the SOM content in the soil. Nijsen et al. (2012) reported that switching from annual food crops to perennial energy crops generally results in an increase in SOM of 1 t C ha-1 yr-1. Degraded soils are generally poor in nutrients and thus perennial crops should be preferred, their being more efficient in capturing nutrients from the soil than annual food crops. Their superiority is owing to their extensive root system and mycorrhizal symbiosis (Nijsen et al. 2012). Crop residues removal or incorporation also play an important role in the nutrient status of the soil, as they are rich in essential plant nutrients (e.g., C, N, P, K, Ca, and Mg).

Table 1. Effect, positive (+) or negative (-), of crops and agricultural practice on soil erosion, soil organic carbon (SOC), soil structure and soil biology

	Soil erosion	SOC	Soil structure	Soil biology
Crops				
Row crops	-	-	-	-
WSGs	+	+	+	+
SRWCs	+	+	+	+
Agricultural practice				
Tillage *with* crop residue removal	-	-	-	-
Tillage *without* crop residue removal	+	+	+	+
No-tillage	+	+	+	+

Their removal directly reduces nutrient pools and alters chemical properties (e.g., pH, electrical conductivity, and cation exchange capacity) (Blanco-Canqui 2010). Degraded soils do not have a favorable soil water regime, and thus the water efficiency of the biofuel plant species is of utmost importance. The water use efficiency of perennial energy crops is generally higher than that of annual food crops, owing to the fact that a) most perennial energy crops have a C4 photosynthetic pathway, and b) more importantly they have larger and deeper root system which allows for higher water absorption efficiency in case of irregular rainfall, and which allows the extraction of water from deeper soil layers (Parrish and Fike 2005).

4.2. Use of Contaminated Soils for Biofuel Production

Some 22 million ha of land worldwide have been degraded by contamination (GACGC 1995). Owing to the increasing industrial development in counties such as China, the area of contaminated soils has probably doubled since 1995. According to Evangelou et al. (2012a) the global area of trace element contaminated soils is approximately 33 million ha. This land can no longer be used for food production but it could be used, with suitble phytomanagement, to produce marketable biomass for energy production. Phytomanagement describes the engineering or manipulation of soil-plant systems to control the fluxes of TEs in the environment. Thus, the goal of phytomanagement is to alleviate deficiencies of essential TEs or to reduce the environmental risk posed by contaminating TEs. A key component of phytomanagement is that it should either cost less than other remediation or fortification technologies, or be a profitable operation, through the production valuable plant biomass (Robinson et al. 2009). The success of phytomanagement depends crucially on the choice of the right plants and cultivation methods. The plants used should minimize contaminant propagation by reducing erosion, and leaching (Robinson et al. 2007), and they should have a low uptake of contaminants (Evangelou et al. 2012a) to ensure safe products. Their cultivation must be practically feasible and economically attractive under the given site and land use conditions (Robinson et al. 2009).

To reduce contaminant leaching plants with high evapotranspiration, such as willow and poplar, should be chosen, as they withdraw transpiration water from the soil through their roots, thus reducing the downward movement of water (Robinson et al. 2003). The choice of plants should also be adapted to the contaminants in the soils. Plants have a propensity to take up certain trace elements (Table 2).

Willow and poplar, in contrast to birch (*Betula pendula*), display a high bioaccumulation factor (trace element concentration in shoot / trace element concentration in soil) for Zn and Cd (Evangelou et al. 2012b; Unterbrunner et al. 2007). Thus, soils with high Zn and Cd concentrations should be planted preferably with birch rather than willow or poplar. The choice of plants is not only important for the mitigation of contaminant propagation but also for the remediation of contaminated soils.

Table 2. Trace element uptake of plants intended for biofuel production

Species	Harvestable part	TE	Bioaccumulation Factor	Reference
Salix caprea	shoot	Cd	2.5	Unterbrunner et al., 2007
		Zn	1.5	
Salix atrocinerea	leaves	Pb	0.29	Reglero et al., 2008
		Zn	3.63	
		Cu	0.25	
		Cd	3.15	
Salix viminalis	wood	Cd	4-5.33	Jensen et al., 2009
		Cu	0.01-0.05	
		Pb	0.002-0.003	
		Zn	0.33-0.58	
Populus tremula	shoot	Cd	2.9	Unterbrunner et al., 2007
		Zn	1.8	
Populus nigra	wood	Cd	1.82	Laureysens et al., 2004
		Zn	0.35	
Betula pendula	wood	Cu	0.03	Rosselli et al., 2003
		Zn	0.32	
		Cd	0.11	
	leaves	Cu	0.01-0.003	Marqui et al., 2007
		Pb	0.01-0.001	
		Zn	1.0-0.01	
Vetiver grass (*Vetiveria zizanioides* L.)	shoot	Pb	0.07	Chiu et al., 2005
		Cu	0.01-0.04	
		Zn	0.03-0.2	
	shoot	Pb	n.d.	Lai and Chen, 2004
		Cd	1.25	
		Zn	0.8	

The effects on microbial community structure and composition in the bulk soil, by the plant species vary, and this aspect subsequently influences the degradation rate of targeted organic pollutants. Leigh et al. (2006) identified Austrian pine (*Populus nigra*) and willow trees as PCB rhizoremediation candidates because of their association with increased numbers of PCB degraders in the root zone as compared to the numbers found for ash (*Fraxinus excelsior*), birch, black locust (*Robinia pseudoacacia*) and non-root-containing soil. According to a study by Ryslava et al. (2003) Tobacco (*Nicotiana tabacum*) sufficiently degraded PCB, whereas black nightshade (*Solanum nigrum*) and alfalfa (*Medicago sativa*) owing to their rhizospheric microbial community displayed significantly lower degradation rates. Ideally therefore phytomanagement of contaminated land would take into account not just the safe and economic use of land but the possibility of improving and perhaps restoring the soil in the long term.

Conclusion

Soil is one of the most undervalued resources we possess. Its degradation is one of the gravest environmental problems we face precisely because its loss in underestimated.

The process of degradation may, in fact, be accelerated by the increasing biofuel production, owing to deforestation, the conversion of grasslands to agricultural land and wrong agricultural practice. Whilst these may bring a higher income in the short run, they can have detrimental economic and environmental effect in the long run.

By a suitable choice of plant species and an appropriate agricultural practice the production of biomass for biofuels has the possibility of not only conserving the environmental quality of soils but even of improving and restoring soils. In general woody crops, such as willow and poplar, and PG, such as switchgrass, should be preferred to agricultural crops such as corn and soybean (Table 1).

With respect to agricultural practice, no-tillage should be preferred to tillage practice because of the decreasing the risk to soil erosion and SOC loss. Moreover, it can improve soil structure and increase both the microbial activity and biomass in the soil. (TWB 2012), (Reglero et al. 2008), (Jensen et al. 2009), (Margui et al. 2007; Rosselli et al. 2003), (Chiu et al. 2005), (Lai and Chen 2004), (Laureysens et al. 2004).

References

Acosta-Martinez V, Mikha MM, Vigil MF (2007) Microbial communities and enzyme activities in soils under alternative crop rotations compared to wheat-fallow for the Central Great Plains. *Appl. Soil Ecol.* 37:41-52.

Al-Kaisi MM, Grote JB (2007) Cropping Systems Effects On Improving Soil Carbon Stocks Of Exposed Subsoil. *Soil Sci. Soc. Am. J.* 71:1381-1388.

Anderson-Teixeira KJ, Davis SC, Masters MD, Delucia EH (2009) Changes in soil organic carbon under biofuel crops. *GCB Bioenergy* 1:75-96.

Bari MA, Schofield NJ (1992) Lowering of a shallow, saline water table by extensive eucalypt reforestation. *Journal of Hydrology* 133:273-291.

Bharati L, Lee KH, Isenhart TM, Schultz RC (2002) Soil-water infiltration under crops, pasture, and established riparian buffer in Midwestern USA. *Agroforestry Systems* 56:249-257.

Blanco-Canqui H (2010) Energy Crops and Their Implications on Soil and Environment. *Agron. J.* 102:403-419.

Blanco-Canqui H, Lal R (2007) Soil and crop response to harvesting corn residues for biofuel production. *Geoderma* 141:355-362.

Blanco-Canqui H, Lal R (2008) Principles of soil conservation and management. *Springer,* Berlin.

Blanco-Canqui H, Lal R (2009a) Corn Stover Removal For Expanded Uses Reduces Soil Fertility And Structural Stability. *Soil Sci. Soc. Am. J.* 73:418-426.

Blanco-Canqui H, Lal R (2009b) Crop Residue Removal Impacts on Soil Productivity and Environmental Quality. *Cr. Rev. Plant Sci.* 28:139-163.

Bridgwater T (2006) Biomass for energy. *J. Sci. Food Agric.* 86:1755-1768.

Bright C (1998) *Life Out of Bounds: Bioinvasion in a Borderless World*. W.W. Norton and Company, New York.

Campbell JE, Lobell DB, Genova RC, Field CB (2008) The global potential of bioenergy on abandoned agriculture lands. *Environ. Sci. Technol.* 42:5791-5794.

Chiu KK, Ye ZH, Wong MH (2005) Enhanced uptake of As, Zn, and Cu by *Vetiveria zizanioides* and *Zea mays* using chelating agents. *Chemosphere* 60:1365-1375.

Congress US (2007) Energy independence and security act. Title II, energy security through increased production of biofuels, subtitle A, renewable fuel standard. In: *Congress US ed. Government Printing Office*, Washington, D.C., USA.

Cornish PM (1989) The Effects of radiata pine plantation establishment and management on water yields and water quality - a review. *Forestry Commission of New South Wales*, Beecroft, Australia.

Demirbas A (2007) Progress and recent trends in biofuels. *Prog. Energy Combust. Sci.* 33:1-18.

Díaz-Zorita M, Perfect E, Grove JH (2002) Disruptive methods for assessing soil structure. *Soil and Tillage Research* 64:3-22.

Doran JW (1987) Microbial biomass and mineralizable nitrogen distributions in no-tillage and plowed soils. *Biol. Fertil. Soils* 5:68-75.

Elustondo J, Angers DA, Laverdiere MR, Ndayegamiye A (1990) Influence of corn croppping and of meadow on aggregation and on organic-matter associated with particle-size fraction of 7 soils. *Can. J. Soil Sci.* 70:395-402.

Evangelou MWH, Conesa HM, Robinson BH, Schulin R (2012a) Biomass production on trace element-contaminated land: a review. *Environmental Engineering Science* 29.

Evangelou MWH, Deram A, Gogos A, Studer B, Schulin R (2012b) Assessment of suitability of tree species for the production of biomass on trace element contaminated soils. *J. Hazard. Mater.* 209–210:233-239.

Fargione J, Hill J, Tilman D, Polasky S, Hawthorne P (2008) Land Clearing and the Biofuel Carbon Debt. *Science* 319:1235-1238.

Friedemann A (2007) Peak Soil: Why cellulosic ethanol, biofuels are unsustainable and a threat to America. *Cultural change*.

GACGC (1995) *World in Transition: The Threat to Soils. Annual Report* - 1994 German Advisory Council on Global Change, Bonn.

Hooker BA, Morris TF, Peters R, Cardon ZG (2005) Long-term Effects Of Tillage And Corn Stalk Return On Soil Carbon Dynamics. *Soil Sci. Soc. Am. J.* 69:188-196.

Hulme PE, Pysek P, Nentwig W, Vila M (2009) Will threat of biological invasions unite the European Union? *Science* 324:40-41.

Jensen JK, Holm PE, Nejrup J, Larsen MB, Borggaard OK (2009) The potential of willow for remediation of heavy metal polluted calcareous urban soils. *Environ. Pollut.* 157:931-937.

Kahle P, Baum C, Boelcke B (2005) Effect of afforestation on soil properties and mycorrhizal formation. *Pedosphere* 15:754-760.

Katsvairo TW, Wright DL, Marois JJ, Hartzog DL, Balkcom KB, Wiatrak PP, Rich JR (2007) Cotton roots, earthworms, and infiltration characteristics in sod-peanut-cotton cropping systems. *Agron. J.* 99:390-398.

Kemper D, Dabney S, Kramer L, Dominick D, Keep T (1992) Hedging against erosion. *J. Soil Water Conserv.* 47:284-288.

Kort J, Collins M, Ditsch D (1998) A review of soil erosion potential associated with biomass crops. *Biomass Bioenerg.* 14:351-359.

Lai HY, Chen ZS (2004) Effects of EDTA on solubility of cadmium, zinc, and lead and their uptake by rainbow pink and vetiver grass. *Chemosphere* 55:421-430.

Lal R, Pimentel D (2007) Biofuels from crop residues. *Soil Tillage Res.* 93:237-238.

Laureysens I, Blust R, De Temmerman L, Lemmens C, Ceulemans R (2004) Clonal variation in heavy metal accumulation and biomass production in a poplar coppice culture: I. Seasonal variation in leaf, wood and bark concentrations. *Environ. Pollut.* 131:485-494.

Leigh MB, Prouzova P, Mackova M, Macek T, Nagle DP, Fletcher JS (2006) Polychlorinated biphenyl (PCB)-degrading bacteria associated with trees in a PCB-contaminated site. *Appl. Environ. Microbiol.* 72:2331-2342.

Ma Z, Wood CW, Bransby DI (2000) Carbon dynamics subsequent to establishment of switchgrass. *Biomass Bioenerg.* 18:93-104.

Margui E, Queralt I, Carvalho ML, Hidalgo M (2007) Assessment of metal availability to vegetation (Betula pendula) in Pb-Zn ore concentrate residues with different features. *Environ. Pollut.* 145:179-184.

Milbrandt A, Overend RP (2009) Assessment of biomass resources from marginal lands in APEC economies - APEC#209-RE-01.4. APEC Energy Working Group.

Munkholm LJ, Hansen EM, Olesen JE (2008) The effect of tillage intensity on soil structure and winter wheat root/shoot growth. *Soil Use and Management* 24:392-400.

Nijsen M, Smeets E, Stehfest E, van Vuuren DP (2012) An evaluation of the global potential of bioenergy production on degraded lands. *GCB Bioenergy* 4:130-147.

OECD/FAO (2011) *OECD-FAO Agricultural Outlook* 2011-2020.

Parliament E, Council (2009) Directive 2009/28/EC of the European Parliament and of the Council of 23 April 2009 on the promotion of the use of energy from renewable sources and amending and subsequently repealing Directives 2001/77/EC and 2003/30/EC. In: Parliament E ed. 2009/28/EC, Brussels.

Parrish DJ, Fike JH (2005) The biology and agronomy of switchgrass for biofuels. *Cr. Rev. Plant Sci.* 24:423-459.

Pimentel D (2006) Soil Erosion: A Food and Environmental Threat. *Environment, Development and Sustainability* 8:119-137.

Pimentel D, Krummel J (1987) Biomass energy and soil erosion: Assessment of resource costs. *Biomass* 14:15-38.

Rachman A, Anderson SH, Gantzer CJ, Alberts EE (2004) Soil hydraulic properties influenced by stiff-grass hedge systems. *Soil Sci. Soc. Am. J.* 68:1386–1393.

Raghu S, Wiedenmann RN, Blouin ML (2006) Biofuels as Invasive Species?University of Arkansas Newswire:http://newswire.uark.edu/article.aspx?id=9161 05.09.

Reglero MM, Monsalve-González L, Taggart MA, Mateo R (2008) Transfer of metals to plants and red deer in an old lead mining area in Spain. *Sci. Total Environ.* 406:287-297.

Righelato R, Spracklen DV (2007) Environment - Carbon mitigation by biofuels or by saving and restoring forests? *Science* 317:902-902.

Robinson B, Green S, Mills T, Clothier B, van der Velde M, Laplane R, Fung L, Deurer M, Hurst S, Thayalakumaran T, van den Dijssel C (2003) Phytoremediation: using plants as biopumps to improve degraded environments. *Aust. J. Soil Res.* 41:599-611.

Robinson BH, Banuelos G, Conesa HM, Evangelou MWH, Schulin R (2009) The phytomanagement of trace elements in soil. *Cr. Rev. Plant Sci.* 28:240-266.

Robinson BH, Green SR, Chancerel B, Mills TM, Clothier BE (2007) Poplar for the phytomanagement of boron contaminated sites. *Environ. Pollut.* 150:225-233.

Rosselli W, Keller C, Boschi K (2003) Phytoextraction capacity of trees growing on a metal contaminated soil. *Plant Soil* 256:265-272.

Ryslava E, Krejcik Z, Macek T, Novakova H, Denmerova K, Mackova M (2003) Study of PCB degradation in real contaminated soil. *Fresenius Environ. Bull.* 12:296-301.

Scott RR, Junyang J (2011) GAIN (Global Agriculultural Information Network) Report Number: 11039 , Biofuels 2011 Annual Report In: *Bean R ed. China - Peoples Republic of, Biofuels Annual Report*. USDA Foreign Agricultural Service.

Simmons BL, Coleman DC (2008) Microbial community response to transition from conventional to conservation tillage in cotton fields. *Appl. Soil Ecol.* 40:518-528.

Spedding TA, Hamel C, Mehuys GR, Madramootoo CA (2004) Soil microbial dynamics in maize-growing soil under different tillage and residue management systems. *Soil Biol. Biochem.* 36:499-512.

Sullivan P (2004) Sustainable soil management: Soil Systems Guide. ATTRA National Sustainable Agriculture Information Service, Fayetteville, Arkansas.

Tilman D, Hill J, Lehman C (2006) Carbon-negative biofuels from low-input high-diversity grassland biomass. *Science* 314:1598-1600.

Treonis AM, Austin EE, Buyer JS, Maul JE, Spicer L, Zasada IA (2010) Effects of organic amendment and tillage on soil microorganisms and microfauna. *Appl. Soil Ecol.* 46:103-110.

TWB (2012) Arable land (hectares per person). The World Bank Group.http://data .worldbank.org/indicator/AG.LND.ARBL.HA.PC Accessed 08 June 2012.

UNEP (2002) State of the Environment and Policy Retrospective: 1972–2002 - Land. In: Clarke R, Lamb R, Ward DR (eds) *GEO - Global Environmental Outlook 3 - Past, present and future perspectives.*, Earthscan Publications Ltd, London, UK, pp 62-89.

UNEP (2012) Part 1: State and Trends of the Environment - Land. In: Ullstein B, de Mattos H, Hawkins C, McMullen CP, Jabbour J, Scharlemann J (eds) *GEO - Global Environmental Outlook 5* - Environment for the future we want., pp 65-96.

Unterbrunner R, Puschenreiter M, Sommer P, Wieshammer G, Tlustos P, Zupan M, Wenzel WW (2007) Heavy metal accumulation in trees growing on contaminated sites in Central Europe. *Environ. Pollut.* 148:107-114.

van Vuuren DP, van Vliet J, Stehfest E (2009) Future bio-energy potential under various natural constraints. *Energ. Policy* 37:4220-4230.

Wilhelm WW, Johnson JME, Karlen DL, Lightle DT (2007) Corn stover to sustain soil organic carbon further constrains Biomass supply. *Agron. J.* 99:1665-1667.

Wilhelm WW, Johnson JMF, Hatfield JL, Voorhees WB, Linden DR (2004) Crop and soil productivity response to corn residue removal: A literature review. *Agron. J.* 96:1-17.

Williams JD, Gollany HT, Siemens MC, Wuest SB, Long DS (2009) Comparison of runoff, soil erosion, and winter wheat yields from no-till and inversion tillage production systems in northeastern Oregon. *J. Soil Water Conserv.* 64:43-52.

Zan CS, Fyles JW, Girouard P, Samson RA (2001) Carbon sequestration in perennial bioenergy, annual corn and uncultivated systems in southern Quebec. *Agric. Ecosyst. Environ.* 86:135-144.

In: Bioenergy Systems, Biological Sources … ISBN: 978-1-62417-330-1
Editor: Michel C. Allard

Chapter 3

EFFECTIVE BIOCATALYSTS FOR LIGNOCELLULOSIC BIOFUEL BY PROTEOMICS STUDY OF FUNGAL SECRETOME

Sunil S. Adav** and Siu Kwan Sze
School of Biological Sciences,
Nanyang Technological University, Singapore

ABSTRACT

Due to the impact of energy crisis and global warming on socioeconomic structure, clean renewable energy is urgently required to replace the fossil energy. Lignocellulosic biomass is the most abundant renewable bioresource that can contribute to significant amount of global energy demand to alleviate global warming. To convert solid biomass to higher value biofuel, the current bottleneck is the hydrolysis of the biomass to monomeric sugar for further fermentation to biofuel. Lignocellulosic biomass is multifarious polymer composed of a complex mixture of cellulose, hemicellulose, and lignin; and its bioconversion to monomeric sugars requires a large number of enzymes. Fungi produce an arsenal of extracellular enzymes, the secretome, that catalyses degradation of complex lignocellulosic biomass. To accelerate the biotechnological application of these enzymes in lignocellulosic biofuel, more recently, proteomics technology has been applied to explore the comprehensive secretory fungal lignocellulolytic enzymes during lignocellulosic biomass degradation. This chapter illustrates the application of state-of-the-art proteomics technology for fungal proteomics, particularly emphasizes on secretory proteome (secretome) and also highlights new insights into the diversity of expressed fungal lignocellulolytic enzymes during cellulose and lignocellulose degradation. A broad understanding of the quantitative composition of extracellular secreted lignocellulolytic enzymes and the underlying synergetic mechanism may shed light on the fundamental understanding and further its application to lignocellulosic biofuel and sustainable development. The lignocellulosic biomass

* Corresponding authors Siu Kwan SZE (Asst Prof.) School of Biological Sciences Nanyang Technological University 60 Nanyang Drive, Singapor 637551 Tel: (65) 6514-1006; Fax: (65) 6791-3856 Email: sksze@ntu.edu.sg

composition, their type, nature and complexity has been correlated with fungal lignocellulolytic enzymes identified by mass spectrometry.

INTRODUCTION

Fungi comprise an important class of microorganisms that make significant contribution to our life and also have significant commercial relevance. Filamentous fungi have a long history in biotechnological industry to produce wide variety of products including antibiotics, specialty chemicals and wide variety of highly active enzymes of industrial interest [1, 2]. In addition, fungi are also important human food, as many mushrooms are edible and different species are cultivated for sale worldwide. Fungi, together with bacteria, recycle major constituents of plant biomass and play major role in biological carbon cycle. Due to their ability to secrete potent lignocellulolytic enzymes, fungi are presumed as potent biomass degrader and hence to explore them and their enzymes for lignocellulosic biorefinery, a strong initiative in sequencing of their genome [3-9] and proteomics identification of expressed proteins has been documented [10-13]. In order to stabilize current atmospheric CO_2 level, minimize green house gas emissions, mitigate global warming and improve climate issues; researchers are exploring fungal lignocellulolytic enzymes for biorefinery e.g. in the production of lignocellulosic biofuels and specialty chemicals.

When growing on plant biomass, they usually grow as long threads called hyphae (figure 1) which interact with substrate, release enzymes and breakdown the complex molecules.

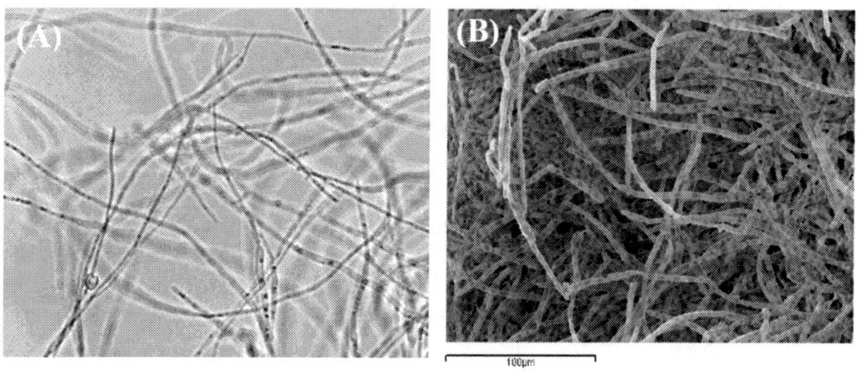

Figure 1. Fungal hyphae under microscope (A) and by scanning electron microscope (B).

Thus, saprophytic fungi use dead organic material for their growth and also convert it into carbon dioxide, and small molecules, such as organic acids. They decompose complex polymers including cellulose (polymer of glucose), hemicellulose (polymer of pentose and hexose sugars) and lignin. On the contrary, there are some fungi that cause diseases in plants and animals, including peoples; and they are generally referred as pathogenic fungi. The plant pathogens causes reduced production or plant death and also affect fruits and vegetables. While, mycorrhizal fungi are mutualist, create a symbiotic relationship (mutually beneficial) with many plants. Thus, fungi are a spoiler of food or as some tasty morsel that can be cooked and eaten. But the major and true purpose of these fungi in nature is recycling of dead organic matter. Fungi together with bacteria decompose complex organic compounds and return minerals into soil and gases to the air, thus making them available for plants and animals. Without these and some other natural recycling processes, lives on the earth remain curtailed.

At present, renewable energy production is on top of global agenda due to global warming, climate change, increasing oil prices and energy security. Lignocellulosic biomass produced through photosynthesis by trapping sunlight energy, offers a compelling solution. Agricultural lignocellulosic wastes, crop residues and forest wastes are renewable, quite abundant in nature and hence considered to be an ideal source of substrates for bioenergy. Again, they are cheaper and widely distributed worldwide. These lignocellulosic biomasses are mainly composed of carbohydrates including cellulose, hemicellulose and pectin; and recalcitrant lignin. The wide variety of crop residues including wheat straw, rice straw, sugar cane bagasse, corn stover, palm, husk etc. have varying amounts of carbohydrate components but lower degree of lignifications than wood. Hence, the bioconversion of these lignocellulosic biomasses, perennial grasses and transgenic fast growing plants etc. into fermentable sugars has potential to produce renewable biofuel and mitigate global warming. But hydrolysis of recalcitrant lignocellulose to its constitutive components is the bottleneck of the process, where lignocellulolytic enzymes play a major role. Therefore, this chapter will merely focus on the lignocellulolytic enzymes for lignocellulosic biofuel, produced by lignocellulose degrading fungal strain *P. chrysosporium*. Since traditional enzyme activity assays using colorimetric methods have several limitations and hence how proteomics technology can be applied to explore comprehensive secretory profile of these fungi will be discussed.

PROTEOMIC TECHNIQUES AND FUNGAL PROTEIN QUANTIFICATION

Proteomics is a member of 'omics' family that gained rapid recognition in disease biomarkers and therapeutics. With the advances in the genome sequencing technology, proteomics is expanding very rapidly in the last decade. Genomics decode all of the genes in an organism while proteomics, in its very simple definition, discover all of the proteins to elucidate their functions. Although, genes are the blueprints of proteins, but comparative analysis of the data produced by gene level studies and information obtained at protein expression level concluded that genetic information often fail to correlate with protein abundance and protein function. The proteomics study usually highlight dynamic nature of the proteome in time-dependent protein expressions [14], physiological stage of microbes [15] or microbial response with environmental conditions. The proteomics technology assures a wide range of potential applications in various stages of drug discovery process, targeted protein identification, toxicological profiling in clinical samples. It's been considered that a global understanding of proteins including their expressions, regulations, interactions, possible post translation modifications etc. at particular disease or environmental condition will reveal the underlying mechanism leading to development of drugs or disease biomarkers. Such literature on microbial protein expressions and their regulation during lignocellulose hydrolysis will direct the possible development in designing enzyme cocktail for lignocellulosic biorefinery. However, little literature on proteomics analysis of microbial lignocellulolytic enzymes is available.

In proteomics, 2D-gel electrophoresis; stable isotope-labeling techniques like isotope-coded affinity tagging (ICAT) [16]; multidimensional liquid chromatography or multidimensional protein identification technology (MudPIT) [17] in combination with on-line ESI-MS/MS provides very reliable relative quantification. An iTRAQ™ labeling technique which is based on stable isotopes allows multiplexing of up to eight different biological samples. In addition to clinical research, this iTRAQ technique has been tested for microbial proteome and secretome profiling [10, 12, 18-20]. Several microbial proteomics studies have applied gel based, gel free and label free quantitation techniques [11, 21-25] to quantify proteome or secretome at particular physiological condition. The iTRAQ reagents which are available as four-plex or eight-plex kits allow multiplexing of up to four or eight different samples in a single LC-MS/MS experiment.

These iTRAQ™ reagents are non-polymeric, isobaric tagging reagents and consist of a reporter group, a balance group, and a peptide reactive group. The peptide reactive group (PRG) covalently links an iTRAQ™ reagent isobaric tag with each lysine side chain and N-terminal group of a peptide. The main purpose of balance group is to ensure an iTRAQ™ reagent-labeled peptide, whether labeled with iTRAQ™ reagent 114, 115, 116, or 117, displays at the same mass [26]. The iTRAQ™ reagent reporter group ions generated during MS/MS fragmentation appear as peaks in the low-mass region between m/z 113 and m/z 121. Since this region is free from other common fragment ions, signals achieved in this region could be due to the reporter ions of corresponding labeled sample digests. Therefore, the relationships can be quantified by comparing the peak area of one reporter group peak to another. The ratio of one peak area to another represents the relative amount of a given peptide in each of the corresponding sample digests. The experimental work flow is presented in figure 2.

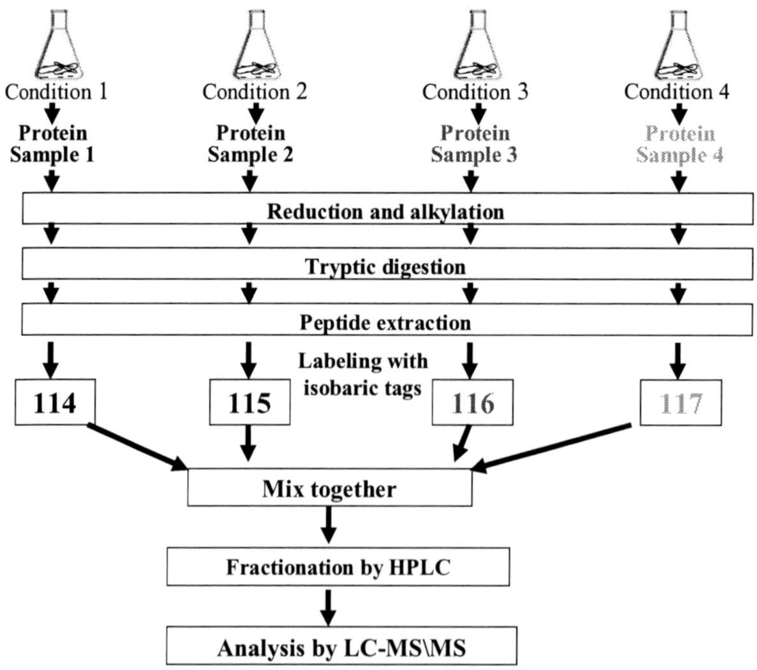

Figure 2. Experimental work flow for iTRAQ quantification of proteins from four different conditions.

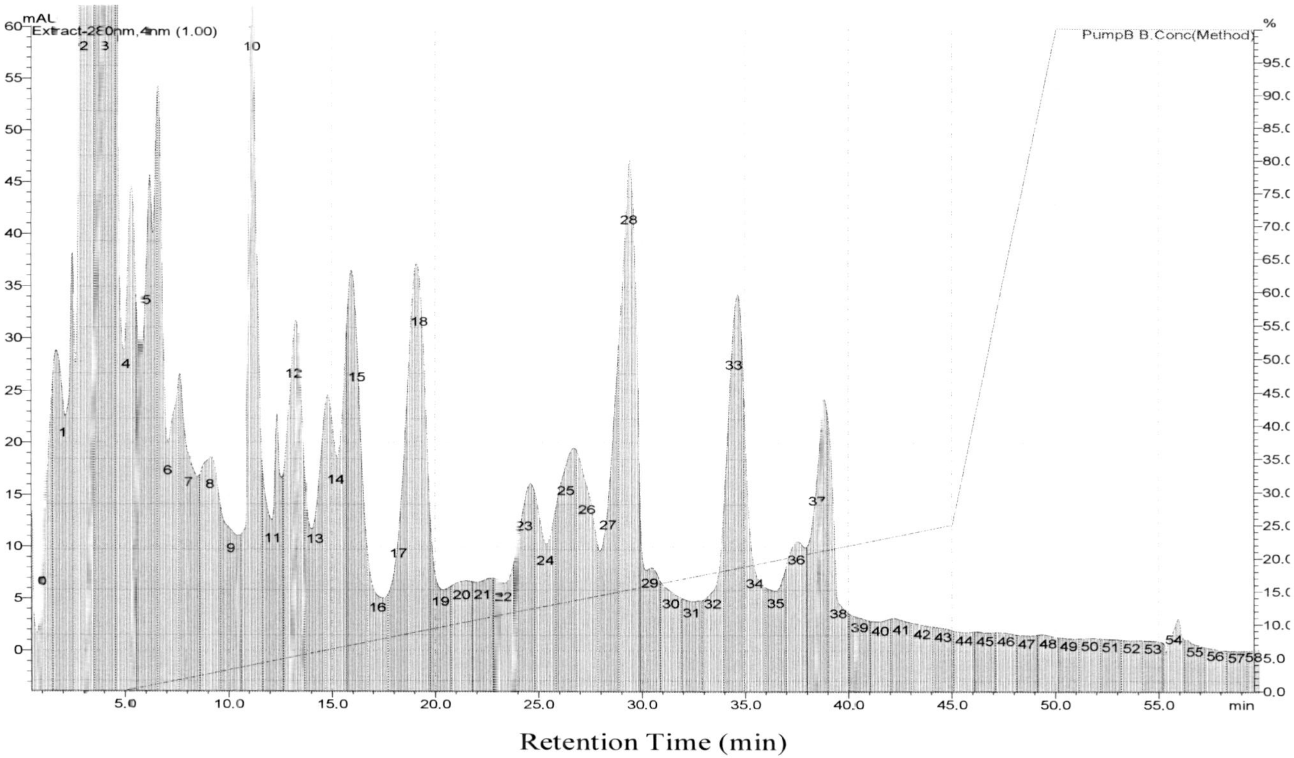

Figure 3. Peptide separation profile achieved by using ERLIC fractionation.

The proteome samples either from higher animals, microbes or plants are highly complex and proteomic methods employing mass spectrometers do not define or identify all proteins in simple one-step process. In proteomics, sample preparation is crucial and necessitates sample simplification either by HPLC-fractionation or separation steps. The main purpose of the sample preparation and fractionation is sample clean up to remove contaminants and enrichment of analytes to improve the quality of analysis or data. The selection of the fractionation method depends on the experimental objective like protein quantification; isoform identification, post-translation modification [27, 28] etc. and may require case-by-case development. Earlier, we optimized salt free electrostatic repulsion-hydrophilic interaction chromatographic (ERLIC) fractionation technique [27] which is very useful in proteomics. Using this earlier reported technique we simplified the secretory protein samples and their peptide separation profile along with the HPLC-gradient is presented in figure 3.

Lignocellulose Biomass and Carbohydrate Hydrolyzing Enzymes

Lignocellulose biomass primarily consist of three major constituents – cellulose, hemicellulose, and lignin; plus some lesser amounts of pectin, protein and other compounds often called extractives. Cellulose comprises between 35 and 50% of the total biomass and its content depends upon the type and species of the plants. In most cases, cellulose fibers are embedded in a matrix of other biopolymers such as hemicellulose and lignin which comprises 20 to 35 and 5 to 30% of plant biomass on dry weight basis. However, the interactions among different polymers depend upon the type and maturity (age) of the plant biomass and are dominating structural features that control the rate and utilization of biomass [29]. The bonding pattern in cellulose divides the cellulose structure into crystalline, a well ordered glucose units bound laterally with hydrogen bonds; and amorphous cellulose which is less ordered [30]. Again, cellulose is a linear chain of several hundred to over thousands of glucose units linked by β-1-4 glycosidic linkages.

The acid/alkali hydrolysis of lignocellulosic biomass generates inhibitory compounds that inhibit microbial growth and also affect down-stream fermentation process. The selective removal of these inhibitory compounds has been found to be very expensive. On the contrary, enzymatic hydrolysis is efficient and has several advantages over acid/alkali pretreatments. The complete hydrolysis of biomass requires different glycoside hydrolases including endo-acting cellulases, exo-acting cellulases and beta-glucosidases.

Several microbial strains have the potential to degrade lignocellulosic biomass. The complex bonding pattern between and among the different components of lignocellulosic biomass such as cellulose, hemicellulose, pectin, lignin and proteins made it difficult to deduce its degradation pathways. Again, degree of lignifications increases with age of the plant and differs with plant species. Of the other several multiple factors that hindered the progress, substrate complexity and the multiplicity of enzymes are the main issues. Hence, irrespective of considerable research on lignocellulose degradation, fundamental questions remain unresolved with regard to mechanisms of its degradation. The efforts to discover novel enzymes and to elucidate lignocellulose degradation mechanism could be highlighted with an increasing number of genomes that have been sequenced so far. About 18 different species including *Aspergillus clavatus* [31] , *Aspergillus flavus, Aspergillus fumigatus* [3, 4], *Aspergillus nidulans* [3], *Aspergillus niger* [5], *Aspergillus oryzae* [4], *Aspergillus terreus, Fusarium graminearum* [6], *Fusarium verticillioides*, *Magnaporhe grisea* [32], *Neurospora crassa* [7], *P. chrysosporium* [9] and *Trichoderma reesei* [8] have been sequenced and annotated; while several projects are in process.

The quantitative expressions of lignocellulolytic proteins by *P. chrysosporium* on pure fibrous cellulose, lignin and mixture of cellulose and lignin have been documented [13]. This protein profile data on simple soluble substrates could be useful in predicting the mechanistic roles of these enzymes in substrate degradation and ascertain synergetic relationships. However, such data could not be useful in establishing efficacy of these hydrolytic enzymes on natural complex lignocellulosic biomasses since enzymatic mechanisms on soluble and insoluble substrates differ significantly [33]. As precisely stated by Zhang et al. [33], no clear relationships exists between cellulase activities on soluble substrate and those on insoluble complex substrates, hence use of soluble substrates for screening cellulases and examining "improved" activities may not be an ideal. Again, expression, efficacy and performance of lignocellulolytic enzymes depends on several factors, including nature and type of substrates; culture conditions, pH, enzyme stability at various conditions; product inhibition and synergism between enzymes. Another major limitation of cellulose hydrolysis by hydrolytic enzymes is its penetration through plant cell wall because cellulose is embedded by matrix of other polymers such as lignin.

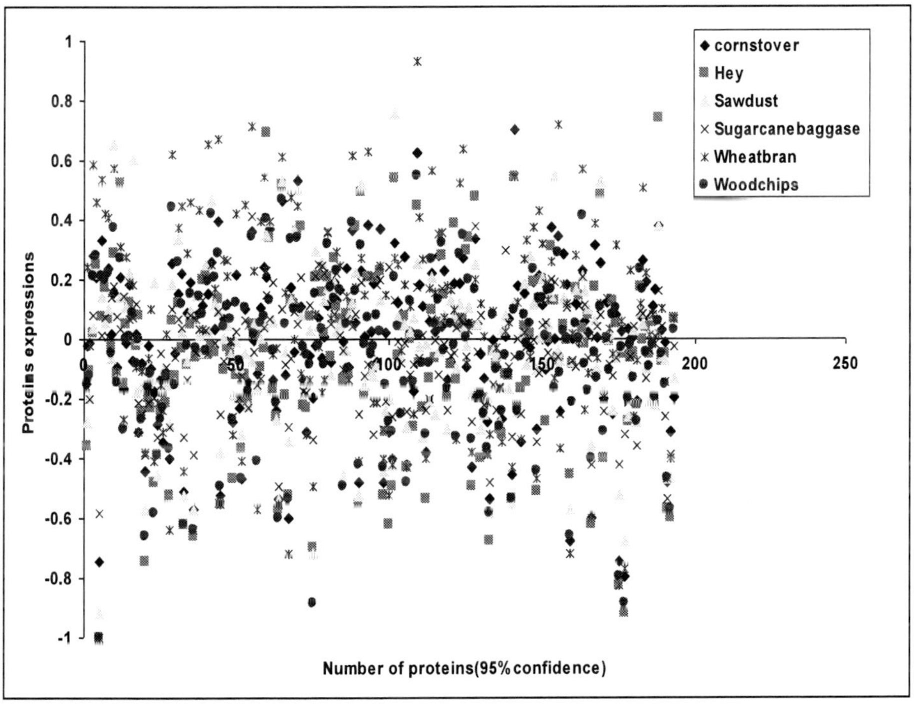

Figure 4. Lignocellulolytic protein expression as a log (iTRAQ ratio) when *P. chrysosporium* was cultivated with different lignocellulosic biomasses. Proteins displayed above the axis indicated up-regulation, below the axis indicated down-regulation while those on the axis were not regulated.

Adav et al. [12], cultivated *P. chrysosporium* with different lignocellulosic biomasses and profiled quantitative expression of extracellular lignocellulolytic enzymes. This study identified more than 300 proteins. Some intracellular proteins were identified indicating minor cell death or lysis during secretome extraction. The expression profile of the secretory proteins after separating them from intracellular proteins using SignalP and transmembrane domain database were presented in figure 4 as a $\log_{(\mathrm{iTRAQ\ ratio})}$ against protein numbers. Protein with positive $\log_{(\mathrm{iTRAQ\ ratio})}$ indicates up-regulation while those with negative values indicates down regulation.

The biological functional classification of these expressed enzymes revealed that the secretory protein of *P. chrysosporium* were a complex mixture of cellulases, glycoside hydrolases, hemicellulases, lignin degrading proteins, proteases, transport and hypothetical proteins. In addition, proteins involved in cell growth, morphogenesis and signaling were also identified. Thus, secretome was constituted with 16.6% cellulases, 19.9% glycoside hydrolases, 15.6% hemicellulases, 10.0% lignin depolymerizing proteins, 19.9% peptidases and proteases, 7.1% transport proteins, 7.1% hypothetical proteins and some lipases, chitinases etc. (Figure 5).

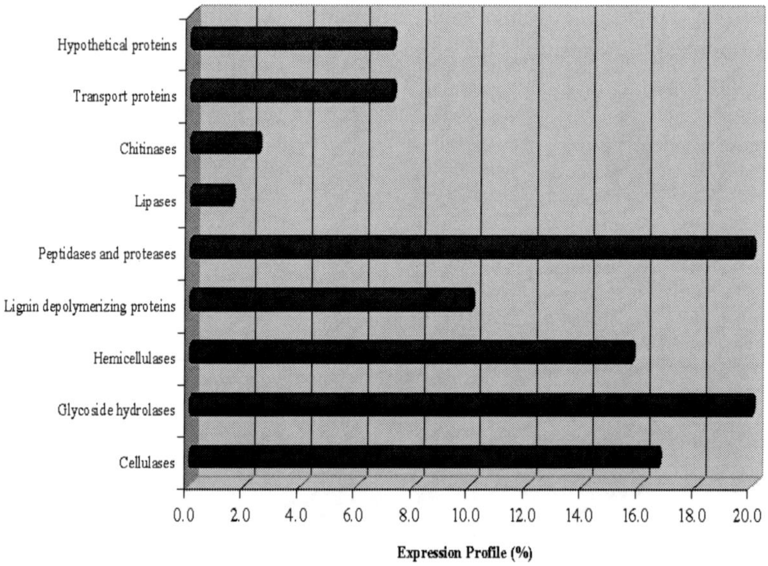

Figure 5. Functional classification of proteins by *P. chrysosporium* when cultivated on different lignocellulosic biomasses.

Further, 68 proteins with cellulose degrading potential were reported. The data analysis revealed that cellulases include the large number of endo- and exo-glucanases which hydrolyze β-1,4-glucosidic bonds of the cellulose chains. It's been assumed that at least three key enzymes are required for complete hydrolysis of cellulose-i) endogucanases catalyses the endohydrolysis of β-1-4 glycosidic linkages creating free chain ends; ii) exoglucanases that cut the dissaccharide cellobiose from the free chain ends and iii) Beta-1,4-glucosidases which hydrolyze the cellobiose and other short cello-oligosaccharides to glucose.

It's well known that endoglucanases alone cannot break down polysaccharides without non-catalytic carbohydrate binding module. Thus, endoglucanses are a complex composed of catalytic domain, cellulose binding domain (CBM) and flexible linker. The CBM facilitate the enzyme by binding the complex to cellulose and thus maintain the proximity of enzyme and the substrate. Thus, CBM can influence the rate of hydrolysis of crystalline cellulose by yet an unknown mechanism. When *P. chrysosporium* was cultivated with different lignocellulosic biomasses, at least seven different types of endoglucanases (figure 6) were expressed and iTRAQ quantified [12]. The CBM undergoes a structural conformation change when the substrate gets fixed in place i.e. active site. Within the active site, glutamine acts as a catalytic nucleophile while tryptophan secures the substrate by hydrophobic interaction at the entrance of the active site tunnel [34, 35]. The enzyme complex is driven along the cellulose micro fibril through energy gained by the hydrolysis of the glycosidic bond of cellulose [36]. Once endoglucanase cleaves the crystalline cellulose in an amorphous zone, exoglucanses acts on these large insoluble chains of cellulose and convert them into smaller chains called cellobiose or cellodextrine. Then, beta-glucosidases hydrolyze soluble cellobiose to glucose which subsequently can be fermented into biofuel. Cellulases system exhibit synergism and four different forms of synergism have been documented (i) endo-exo synergy between endoglucanases and exoglucanases, (ii) exo-exo synergy between exoglucanases processing from the reducing and non-reducing ends of cellulose chains, (iii) synergy between exoglucanases and beta-glucosidases that remove cellobiose (and cellodextrins) as end products of the first two enzymes, and (iv) intra-molecular synergy between catalytic domains and CBMs [37, 38].

Fungal cellulases that dominate the industrial application are under intensive investigation [39]. *P. chrysosporium* has high potential industrial application. According to genome sequence, *P. chrysosporium* encodes seven CBH1, one CBH2, two EG1, two EG2, one EG3 and fourteen EG4 [8].

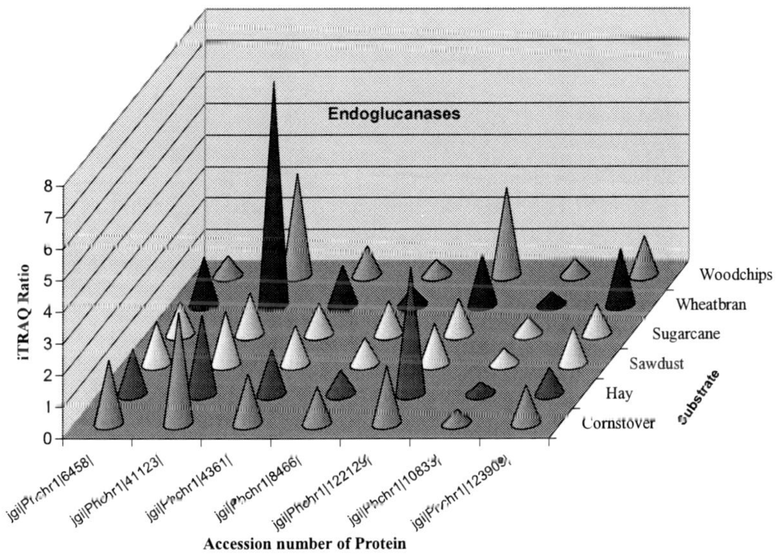

Figure 6. Expression profile of endoglucanases when *P. chrysosporium* was cultured with different lignocellulosic biomasses. [GH5 Endoglucanase (jgi|Phchr1|6458|), Endoglucanase B (jgi|Phchr1|41123|), GH5 Endoglucanase (jgi|Phchr1|4361|), GH12 Endoglucanase (jgi|Phchr1|8466|), Endoglucanase B (jgi|Phchr1|122129|), GH16 Laminarinase; Endo-1,3(4)-Beta- Glucanase (jgi|Phchr1|10833|), GH16 Laminarinase; Endo-1,3(4)-Beta-Glucanase (jgi|Phchr1|123909|)].

When *P. chrysosporium* was cultivated with different lignocellulosic biomasses, at least seven different types of endoglucanases and five different exoglucanases were expressed and iTRAQ quantified. (figure 6, figure 7). Glycoside hydrolases (or glycosyl hydrolases) catalyze the hydrolysis of the glycosidic linkage between two or more carbohydrates or between a carbohydrate and a non-carbohydrate moiety. Based on the amino acid sequence similarities, these enzymes were classified into GH1 to GH130 families. Thus, our study on the lignocellulolytic enzymes by *P. chrysosporium* during lignocellulose utilization found upregulation of GH6, GH7 exoglucanases and GH5, GH12, GH16, GH61 endoglucanases. This upregulation suggests that these enzymes were involved in the cellulosic biomass degradation. While novel GH1, GH3, GH31, GH55 beta-, alpha-glucosidases were also upregulated in tested substrates.

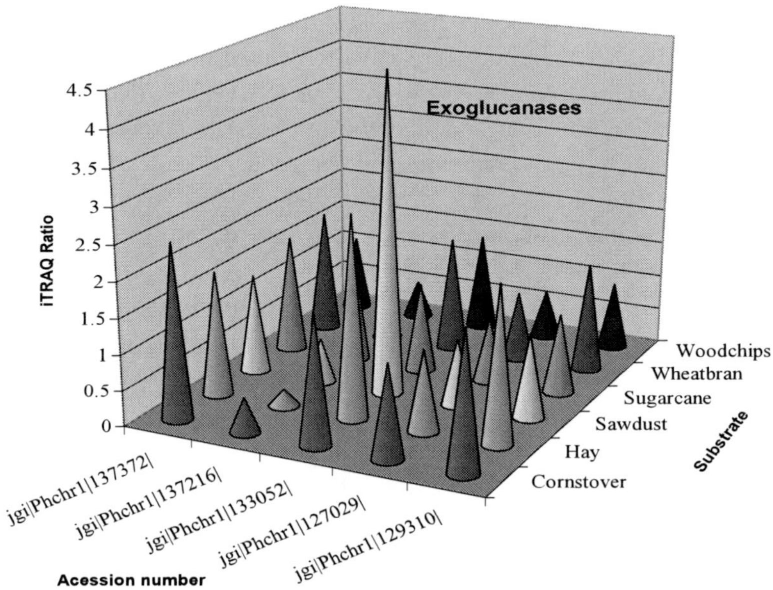

Figure 7. Expression profile of exoglucanases when *P. chrysosporium* was cultured with different lignocellulosic biomasses. [GH7 Cellulase, Cellobiohydrolase Cel7d (jgi|Phchr1|137372|), GH7 Exoglucanase 1 (jgi|Phchr1|137216|), GH6 Exocellobiohydrolase (jgi|Phchr1|133052|), GH7 Exo-cellobiohydrolase; Exoglucanase 1, (jgi|Phchr1|127029|) Exo-cellobiohydrolase I (jgi|Phchr1|129310|)].

In addition to cellobiohydrolases and possible endoglucanases, *P. chrysosporium* also produces cellobiose dehydrogenase (CDH) that, in the presence of O_2, oxidizes cellobiose to cellobionolactone. When *P. chrysosporium* was cultivated using different lignocellulosic biomass, CDH was abundantly produced and iTRAQ quantification revealed its up-regulation. According to Henriksson et al. [40-42], CDH is an extracellular redox enzyme of ping-pong type that generates hydroxyl radicals by reducing Fe^{3+} to Fe^{2+} and O_2 to H_2O_2 and play important role in lignin degradation. Thus, CDH can play major role in both cellulose and lignin degradation.

Beta-glucosidase, an important glucosidase enzyme that acts upon β-1-4 bonds linking two glucose or glucose-substituted molecules (i.e., the disaccharide cellobiose). It is an exocellulase with specificity for a variety of beta-D-glycoside substrates. It's key enzyme in cellulose hydrolysis that regulates this pathway.

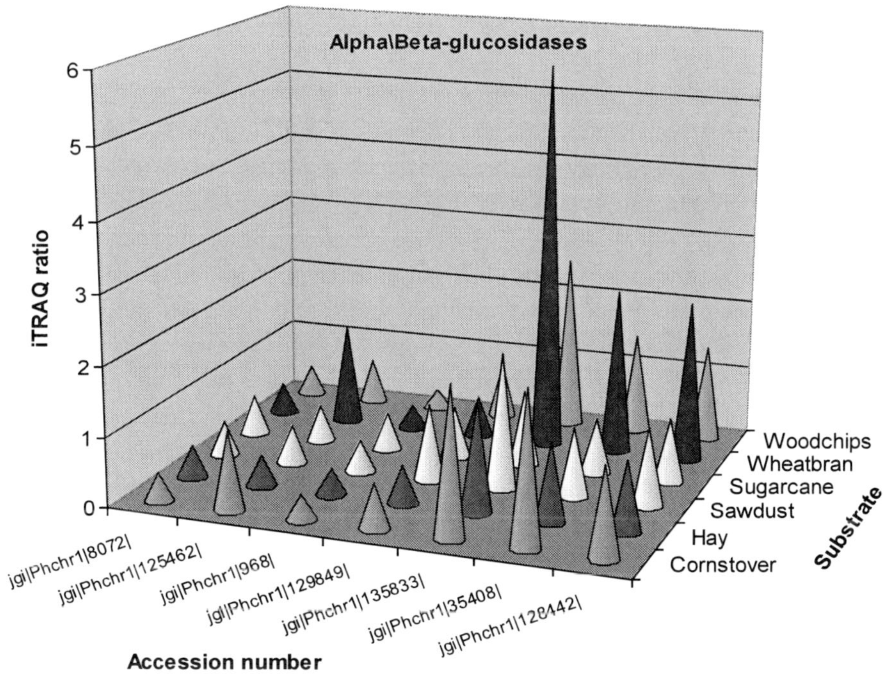

Figure 8. Glucosidases expressed during cultivation of *P. chrysosporium* with different lignocellulosic biomass. [Glucan 1,3-beta-glucosidase (jgi|Phchr1|8072|), Alpha-glucosidase precursor (jgi|Phchr1|125462|), Alpha-glucosidase (jgi|Phchr1|968|), Beta-glucosidase (jgi|Phchr1|129849|), Alpha-glucosidase (jgi|Phchr1|135833|), Alpha-glucosidase (jgi|Phchr1|35408|), Beta-glucosidase (jgi|Phchr1|128442|)].

When *P. chrysosporium* was grown with different lignocellulosic biomasses, several different glucosidases were quantified and their comparative expressions were presented in figure 8. It catalyzes the hydrolysis of terminal non-reducing residues in beta-D-glucosides with release of glucose.

Hemicellulose, the second most abundant renewable biomass represents about 15–35% of plant biomass. The main enzymes needed for hemicellulose degradation are xylanases, assisted by accessory enzymes such as xylosidases, glucuronidase, arabinofuranosidase and acetylxylan esterase making the xylan backbone more accessible [43]. Other enzymes such as mannanase, and mannosidase, carboxyesterases etc. are also required for complete hemicelluloses hydrolysis. The hemicellulose degrading xylanase, endo-polygalacturonase PG1, rhamnogalacturonase, beta-xylosidase, alpha-glucuronidase, ferulic and p coumaric esterases, alpha-L-rabinofuranosidase, endo-arabinase, carboxyesterase, acetylxylan esterase etc. were expressed [12] when *P. chrysosporium* was cultivated with different lignocellulosic biomasses indicating its potential to hydrolyze hemicellulosic component of lignocellulosic biomass. The secretome analysis also revealed the secretion of mannanase Man5C, beta-1,3-mannanase and CEL4a mannanase suggesting mannan hydrolysis potential of *P. chrysosporium*. The lignocellulolytic enzymes by *T. reesei* on saw dust have been documented [18].

Lignin Degrading Enzymes by *P. Chrysosporium*

The main lignin depolymerizing enzymes are lignin peroxidase (LiP), manganese-peroxidase (MnP) and laccases. Since these enzymes are too big to penetrate plant cell wall, hence microbes initially secrete several high diffusing oxidases, reactive radical generating enzymes, and quinine reducing enzymes [44]. The iTRAQ-quntification of lignin degrading enzymes such as copper radical oxidase, cellobiose dehydrogenase, glucose oxidase, isoamyl alcohol oxidase, peroxiredoxins, pyranose 2-oxidase, quinone oxidoreductase, iron-containing alcohol dehydrogenase and other redox enzymes indicated lignin degrading ability of *P. chrysosporium* and is consistent with earlier reports [45-49]. Several oxidases play a major role in generation of peroxide which is essential for peroxidative reactions. In our study, five different copper radical oxidases were quantified when this fungal strain was cultured with different lignocellulosic biomasses. *P. chrysosporium* genome revealed six copper radical oxidase (cro) genes that cluster within a larger cluster of LiP genes and genes cro3, cro4 and cro5 show conservation of sequence. Wymelenberg et al. [50] identified six encoding genes (cro1–cro6), while based on their position Martinez et al. [9] emphasized potential synergizm between peroxidases and copper radical oxidases.

The involvement of CDH in generation of highly reactive hydroxyl radicals that participate in lignocellulose depolymerization has been documented [41, 42]. When *P. chrysosporium* was grown with different lignocellulosic biomasses, CDH was significantly upregulated. This enzyme plays major role in lignin degradation by (i) breaking beta-ethers, demethoxylating aromatic structures in lignin, and (iii) introducing hydroxyl groups in non-phenolic lignin [41, 42, 51]. CDH is involved in both lignin and cellulose hydrolysis; and can be converted into cellobiose-quinone oxidoreductase (CBQ) using proteolytic cleavage [52] or by proteases secreted by *P. chrysosporium*. The detailed characterization of CDH could help to tailor lignin degradation mechanism. Although pyranose 2-oxidase is preferentially localized in the periplasmic space and the associated membranous but experimental evidence supports it's important role in wood decay [53]. Again, its transcript patterns have been shown similar to lignin peroxidases and glyoxal oxidase, supporting a role in lignocellulose degradation [54]. The exact role of other oxidases such as glucose oxidase and alcohol oxidases remain uncertain, but these enzymes support a redox cycle through peroxide [47, 55]. Thus, our study revealed the expression of different peroxide-generating and electron-transfer enzymes by *P. chrysosporium* when cultivated with lignocellulosic biomass. The precise role of multiple copper radical oxidases and FAD oxidoreductases in lignocellulose degradation remains to be established.

In the process of lignin degradation, the cleavage of β-aryl ether is the crucial step, because this intermolecular linkage is the most abundant in lignin. During lignocellulose utilization, high level expression of glutathione S-transferase highlights its potential role in lignocellulose degradation. According to Otsuka et al. [56] and other researchers [57, 58] glutathione S-transferase catalyzes the specific hydrolytic cleavage of lignin guaiacyl-glycerol beta-aryl ether linkages. Masai et al. [57] characterized three GST genes, ligF, ligE, and ligG from lignin degrading *Sphingomonas paucimobilis* SYK-6, and described their role in beta-aryl ether cleavage.

SUMMARY AND PROSPECTIVE

Functional genomics techniques, such as proteomics as well as transcriptomics, provide a new understanding of microbial responses in biomass degradation. Proteomics evaluation of cellular or physiological response to different substrates and environmental adaptation of microorganisms is particularly rewarding because not a few proteins but majority of the proteins involved in the pathways are simultaneously analyzed. Thus, proteomics profiling of lignocellulose degrading enzymes of different lignocellulolytic microbial strains could be perfect screening tool that will highlight expressions and regulation of potent lignocellulolytic enzymes. Within the group of functional genomics techniques, proteomics will keep its central position because it deals with the main candidate player, with the proteins. In lignocellulose degradation, the secreted extracellular enzymes are the key players in the degradation process and it's difficult to study them at transcriptional level. Proteomics reveals the proteins involved in particular process, but to fully understand the detailed mechanism/function, these candidate proteins need to be further characterized. These screened and characterized potent lignocellulolytic proteins can be considered for designing lignocellulose degrading enzyme cocktail.

The function of the *P. chrysosporium* genome can be explored by proteomics technique for the quantitative expressions of enzymes during different lignocellulosic biomass degradation. This represents important advances in the molecular genetics of Basidiomycetes. This chapter provides an example of application of proteomics technique in quantitative profiling of expressed lignocellulolytic proteins for delineating the molecular pathways in carbon recycling. In addition to lignin degradation potential of *P. chrysosporium*, quantitative proteomic analysis of its secretory proteins during lignocellulose biomass hydrolysis highlighted its cellulose and hemicellulose degrading enzymes. Through iTRAQ-quantification of secretory proteins, the existence of many enzymes, particularly peroxidases and their potential complex relationship and synergistic effect was reveled. The proteomics result provides valuable biological insight and potential mechanism of biomass degradation. However, the detailed function of proteins and the exact role need to be further elucidated.

REFERENCES

[1] Punt, P. J., Levasseur, A., Visser, H., Wery, J., and Record, E. (2011) Fungal protein production: Design and production of chimeric proteins. pp. 57-69.

[2] Kniemeyer, O. (2011) Proteomics of eukaryotic microorganisms: The medically and biotechnologically important fungal genus *Aspergillus*. *Proteomics* 11, 3232-3243.

[3] Galagan, J. E., Calvo, S. E., Cuomo, C., Ma, L. J., Wortman, J. R., Batzoglou, S., Lee, S. I., Baştürkmen, M., Spevak, C. C., Clutterbuck, J., Kapitonov, V., Jurka, J., Scazzocchio, C., Farman, M., Butler, J., Purcell, S., Harris, S., Braus, G. H., Draht, O., Busch, S., D'Enfert, C., Bouchier, C., Goldman, G. H., Bell-Pedersen, D., Griffiths-Jones, S., Doonan, J. H., Yu, J., Vienken, K., Pain, A., Freitag, M., Selker, E. U., Archer, D. B., Peñalva, M. Á., Oakley, B. R., Momany, M., Tanaka, T., Kumagai, T., Asai, K., Machida, M., Nierman, W. C., Denning, D. W., Caddick, M., Hynes, M., Paoletti, M., Fischer, R., Miller, B., Dyer, P., Sachs, M. S., Osmani, S. A., and Birren,

B. W. (2005) Sequencing of *Aspergillus nidulans* and comparative analysis with *A. fumigatus* and *A. oryzae*. *Nature* 438, 1105-1115.

[4] Machida, M., Asai, K., Sano, M., Tanaka, T., Kumagai, T., Terai, G., Kusumoto, K. I., Arima, T., Akita, O., Kashiwagi, Y., Abe, K., Gomi, K., Horiuchi, H., Kitamoto, K., Kobayashi, T., Takeuchi, M., Denning, D. W., Galagan, J. E., Nierman, W. C., Yu, J., Archer, D. B., Bennett, J. W., Bhatnagar, D., Cleveland, T. E., Fedorova, N. D., Gotoh, O., Horikawa, H., Hosoyama, A., Ichinomiya, M., Igarashi, R., Iwashita, K., Juvvadi, P. R., Kato, M., Kato, Y., Kin, T., Kokubun, A., Maeda, H., Maeyama, N., Maruyama, J. I., Nagasaki, H., Nakajima, T., Oda, K., Okada, K., Paulsen, I., Sakamoto, K., Sawano, T., Takahashi, M., Takase, K., Terabayashi, Y., Wortman, J. R., Yamada, O., Yamagata, Y., Anazawa, H., Hata, Y., Koide, Y., Komori, T., Koyama, Y., Minetoki, T., Suharnan, S., Tanaka, A., Isono, K., Kuhara, S., Ogasawara, N., and Kikuchi, H. (2005) Genome sequencing and analysis of *Aspergillus oryzae*. *Nature* 438, 1157-1161.

[5] Pel, H. J., De Winde, J. H., Archer, D. B., Dyer, P. S., Hofmann, G., Schaap, P. J., Turner, G., De Vries, R. P., Albang, R., Albermann, K., Andersen, M. R., Bendtsen, J. D., Benen, J. A. E., Van Den Berg, M., Breestraat, S., Caddick, M. X., Contreras, R., Cornell, M., Coutinho, P. M., Danchin, E. G. J., Debets, A. J. M., Dekker, P., Van Dijck, P. W. M., Van Dijk, A., Dijkhuizen, L., Driessen, A. J. M., D'Enfert, C., Geysens, S., Goosen, C., Groot, G. S. P., De Groot, P. W. J., Guillemette, T., Henrissat, B., Herweijer, M., Van Den Hombergh, J. P. T. W., Van Den Hondel, C. A. M. J. J., Van Der Heijden, R. T. J. M., Van Der Kaaij, R. M., Klis, F. M., Kools, H. J., Kubicek, C. P., Van Kuyk, P. A., Lauber, J., Lu, X., Van Der Maarel, M. J. E. C., Meulenberg, R., Menke, H., Mortimer, M. A., Nielsen, J., Oliver, S. G., Olsthoorn, M., Pal, K., Van Peij, N. N. M. E., Ram, A. F. J., Rinas, U., Roubos, J. A., Sagt, C. M. J., Schmoll, M., Sun, J., Ussery, D., Varga, J., Vervecken, W., Van De Vondervoort, P. J. J., Wedler, H., Wösten, H. A. B., Zeng, A. P., Van Ooyen, A. J. J., Visser, J., and Stam, H. (2007) Genome sequencing and analysis of the versatile cell factory *Aspergillus niger* CBS 513.88. *Nature Biotechnology* 25, 221-231.

[6] Fungal genome initiative. Broad Institute of Harvard and MIT (http://www.broad.mit.edu).

[7] Galagan, J. E., Calvo, S. E., Borkovich, K. A., Selker, E. U., Read, N. O., Jaffe, D., FitzHugh, W., Ma, L. J., Smirnov, S., Purcell, S., Rehman, B., Elkins, T., Engels, R., Wang, S., Nielsen, C. B., Butler, J., Endrizzi, M., Qui, D., Ianakiev, P., Bell-Pedersen, D., Nelson, M. A., Werner-Washburne, M., Selitrennikoff, C. P., Kinsey, J. A., Braun, E. L., Zelter, A., Schulte, U., Kothe, G. O., Jedd, G., Mewes, W., Staben, C., Marcotte, E., Greenberg, D., Roy, A., Foley, K., Naylor, J., Stange-Thomann, N., Barrett, R., Gnerre, S., Kamal, M., Kamvysselis, M., Mauceli, E., Bielke, C., Rudd, S., Frishman, D., Krystofova, S., Rasmussen, C., Metzenberg, R. L., Perkins, D. D., Kroken, S., Cogoni, C., Macino, G., Catcheside, D., Li, W., Pratt, R. J., Osmani, S. A., DeSouza, C. P. C., Glass, L., Orbach, M. J., Berglund, J. A., Voelker, R., Yarden, O., Plamann, M., Seiler, S., Dunlap, J., Radford, A., Aramayo, R., Natvig, D. O., Alex, L. A., Mannhaupt, G., Ebbole, D. J., Freitag, M., Paulsen, I., Sachs, M. S., Lander, E. S., Nusbaum, C., and Birren, B. (2003) The genome sequence of the filamentous fungus *Neurospora crassa*. *Nature* 422, 859-868.

[8] Martinez, D., Berka, R. M., Henrissat, B., Saloheimo, M., Arvas, M., Baker, S. E., Chapman, J., Chertkov, O., Coutinho, P. M., Cullen, D., Danchin, E. G. J., Grigoriev, I. V., Harris, P., Jackson, M., Kubicek, C. P., Han, C. S., Ho, I., Larrondo, L. F., De Leon, A. L., Magnuson, J. K., Merino, S., Misra, M., Nelson, B., Putnam, N., Robbertse, B., Salamov, A. A., Schmoll, M., Terry, A., Thayer, N., Westerholm-Parvinen, A., Schoch, C. L., Yao, J., Barbote, R., Nelson, M. A., Detter, C., Bruce, D., Kuske, C. R., Xie, G., Richardson, P., Rokhsar, D. S., Lucas, S. M., Rubin, E. M., Dunn-Coleman, N., Ward, M., and Brettin, T. S. (2008) Genome sequencing and analysis of the biomass-degrading fungus *Trichoderma reesei* (syn. Hypocrea jecorina). *Nature Biotechnology* 26, 553-560.

[9] Martinez, D., Larrondo, L. F., Putnam, N., Sollewijn Gelpke, M. D., Huang, K., Chapman, J., Helfenbein, K. G., Ramaiya, P., Detter, J. C., Larimer, F., Coutinho, P. M., Henrissat, B., Berka, R., Cullen, D., and Rokhsar, D. (2004) Genome sequence of the lignocellulose degrading fungus *Phanerochaete chrysosporium* strain RP78. *Nature Biotechnology* 22, 695-700.

[10] Adav, S. S., Li, A. A., Manavalan, A., Punt, P., and Sze, S. K. (2010) Quantitative iTRAQ secretome analysis of *Aspergillus niger* reveals novel hydrolytic enzymes. *Journal of Proteome Research* 9, 3932-3940.

[11] Adav, S. S., Ravindran, A., Chao, L. T., Tan, L., Singh, S., and Sze, S. K. (2011) Proteomic analysis of pH and strains dependent protein secretion of *Trichoderma reesei*. *Journal of Proteome Research* 10, 4579-4596.

[12] Adav, S. S., Ravindran, A., and Sze, S. K. (2012) Quantitative proteomic analysis of lignocellulolytic enzymes by *Phanerochaete chrysosporium* on different lignocellulosic biomass. *Journal of Proteomics* 75, 1493-1504.

[13] Manavalan, A., Adav, S. S., and Sze, S. K. (2011) ITRAQ-based quantitative secretome analysis of *Phanerochaete chrysosporium*. *Journal of Proteomics* 75, 642-654.

[14] Li, X., Arslan, F., Ren, Y., Adav, S. S., Poh, K. K., Sorokin, V., Lee, C. N., De Kleijn, D., Lim, S. K., and Sze, S. K. (2012) Metabolic adaptation to a disruption in oxygen supply during myocardial ischemia and reperfusion is underpinned by temporal and quantitative changes in the cardiac proteome. *Journal of Proteome Research* 11, 2331-2346.

[15] Hecker, M., and Völker, U. (2004) Towards a comprehensive understanding of *Bacillus subtilis* cell physiology by physiological proteomics. *Proteomics* 4, 3727-3750.

[16] Gygi, S. P., Rist, B., Gerber, S. A., Turecek, F., Gelb, M. H., and Aebersold, R. (1999) Quantitative analysis of complex protein mixtures using isotope-coded affinity tags. *Nature Biotechnology* 17, 994-999.

[17] Washburn, M. P., Wolters, D., and Yates, J. R. (2001) Large-scale analysis of the yeast proteome by multidimensional protein identification technology. *Nature Biotechnology* 19, 242-247.

[18] Adav, S. S., Chao, L. T., and Sze, S. K. (2012) Quantitative secretomic analysis of *Trichoderma reesei* strains reveals enzymatic composition for lignocellulosic biomass degradation. *Molecular and Cellular Proteomics*, doi:10.1074/mcp.M1111.012419.

[19] Adav, S. S., Ng, C. S., Arulmani, M., and Sze, S. K. (2010) Quantitative iTRAQ secretome analysis of cellulolytic *Thermobifida fusca*. *Journal of Proteome Research* 9, 3016-3024.

[20] Adav, S. S., Ng, C. S., and Sze, S. K. (2011) ITRAQ-based quantitative proteomic analysis of *Thermobifida fusca* reveals metabolic pathways of cellulose utilization. *Journal of Proteomics* 74, 2112-2122.
[21] Adav, S. S., Cheow, E. S. H., Ravindran, A., Dutta, B., and Sze, S. K. (2012) Label free quantitative proteomic analysis of secretome by *Thermobifida fusca* on different lignocellulosic biomass *Journal of Proteomics,*, 75, 3694-3706.
[22] Wongwilaiwalin, S., Rattanachomsri, U., Laothanachareon, T., Eurwilaichitr, L., Igarashi, Y., and Champreda, V. (2010) Analysis of a thermophilic lignocellulose degrading microbial consortium and multi-species lignocellulolytic enzyme system. *Enzyme and Microbial Technology* 47, 283-290.
[23] Ryu, J. S., Shary, S., Houtman, C. J., Panisko, E. A., Korripally, P., St. John, F. J., Crooks, C., Siika-aho, M., Magnuson, J. K., and Hammel, K. E. (2011) Proteomic and functional analysis of the cellulase system expressed by *Postia placenta* during brown rot of solid wood. *Applied and Environmental Microbiology* 77, 7933-7941.
[24] Dunne, J. C., Li, D., Kelly, W. J., Leahy, S. C., Bond, J. J., Attwood, G. T., and Jordan, T. W. (2012) Extracellular polysaccharide-degrading proteome of *Butyrivibrio proteoclasticus*. *Journal of Proteome Research* 11, 131-142.
[25] Bouws, H., Wattenberg, A., and Zorn, H. (2008) Fungal secretomes - Nature's toolbox for white biotechnology. *Applied Microbiology and Biotechnology* 80, 381-388.
[26] Burkhart, J. M., Vaudel, M., Zahedi, R. P., Martens, L., and Sickmann, A. (2011) iTRAQ protein quantification: A quality-controlled workflow. *Proteomics* 11, 1125-1134.
[27] Hao, P., Guo, T., Li, X., Adav, S. S., Yang, J., Wei, M., and Sze, S. K. (2010) Novel application of electrostatic repulsion-hydrophilic interaction chromatography (ERLIC) in shotgun proteomics: Comprehensive profiling of rat kidney proteome. *Journal of Proteome Research* 9, 3520-3526.
[28] Hao, P., Qian, J., Dutta, B., Cheow, E. S. H., Sim, K. H., Meng, W., Adav, S. S., Alpert, A., and Sze, S. K. (2012) Enhanced separation and characterization of deamidated peptides with RP-ERLIC-based multidimensional chromatography coupled with tandem mass spectrometry. *Journal of Proteome Research* 11, 1804-1811.
[29] Lynd, L. R., Weimer, P. J., Van Zyl, W. H., and Pretorius, I. S. (2002) Microbial cellulose utilization: Fundamentals and biotechnology. *Microbiology and Molecular Biology Reviews* 66, 506-577.
[30] Jacobsen, S. E., and Wyman, C. E. (2000) Cellulose and hemicellulose hydrolysis models for application to current and novel pretreatment processes. *Applied Biochemistry and Biotechnology - Part A Enzyme Engineering and Biotechnology* 84-86, 81-96.
[31] Wortman, J. R., Fedorova, N., Crabtree, J., Joardar, V., Maiti, R., Haas, B. J., Amedeo, P., Lee, E., Angiuoli, S. V., Jiang, B., Anderson, M. J., Denning, D. W., White, O. R., and Nierman, W. C. (2006) Whole genome comparison of the *A. fumigatus* family. *Medical Mycology* 44, 3-7.
[32] Dean, R. A., Talbot, N. J., Ebbole, D. J., Farman, M. L., Mitchell, T. K., Orbach, M. J., Thon, M., Kulkarni, R., Xu, J. R., Pan, H., Read, N. D., Lee, Y. I., Carbone, I., Brown, D., Yeon, Y. O., Donofrio, N., Jun, S. J., Soanes, D. M., Djonovic, S., Kolomlots, E., Rehmeyer, C., Li, W., Harding, M., Kim, S., Lebrun, M. H., Bohnert, H., Coughlan, S., Butler, J., Calvo, S., Ma, L. J., Nicol, R., Purcell, S., Nusbaum, C., Galagan, J. E., and

Dirren, B. W. (2005) The genome sequence of the rice blast fungus *Magnaporthe grisea*. *Nature* 434, 980-986.

[33] Percival Zhang, Y. H., Himmel, M. E., and Mielenz, J. R. (2006) Outlook for cellulase improvement: Screening and selection strategies. *Biotechnology Advances* 24, 452-481.

[34] Zhao, X., Rignall, T. R., McCabe, C., Adney, W. S., and Himmel, M. E. (2008) Molecular simulation evidence for processive motion of *Trichoderma reesei* Cel7A during cellulose depolymerization. *Chemical Physics Letters* 460, 284-288.

[35] Nimlos, M. R., Matthews, J. F., Crowley, M. F., Walker, R. C., Chukkapalli, G., Brady, J. W., Adney, W. S., Cleary, J. M., Zhong, L., and Himmel, M. E. (2007) Molecular modeling suggests induced fit of Family I carbohydrate-binding modules with a broken-chain cellulose surface. *Protein Engineering, Design and Selection* 20, 179-187.

[36] Igarashi, K., Koivula, A., Wada, M., Kimura, S., Penttilä, M., and Samejima, M. (2009) High speed atomic force microscopy visualizes processive movement of *Trichoderma reesei* cellobiohydrolase I on crystalline cellulose. *Journal of Biological Chemistry* 284, 36186-36190.

[37] Teeri, T. T. (1997) Crystalline cellulose degradation: New insight into the function of cellobiohydrolases. *Trends in Biotechnology* 15, 160-167.

[38] Din, N., Damude, H. G., Gilkes, N. R., Miller Jr, R. C., Warren, R. A. J., and Kilburn, D. G. (1994) C1-C(x) revisited: Intramolecular synergism in a cellulase. *Proceedings of the National Academy of Sciences of the United States of America* 91, 11383-11387.

[39] Gusakov, A. V., Sinitsyn, A. P., Manenkova, J. A., and Protas, O. V. (1992) Enzymatic saccharification of industrial and agricultural lignocellulosic wastes - Main features of the process. *Applied Biochemistry and Biotechnology* 34-35, 625-637.

[40] Henriksson, G., Pettersson, G., Johansson, G., Ruiz, A., and Uzcategui, E. (1991) Cellobiose oxidase from *Phanerochaete chrysosporium* can be cleaved by papain into two domains. *European Journal of Biochemistry* 196, 101-106.

[41] Henriksson, G., Johansson, G., and Pettersson, G. (2000) A critical review of cellobiose dehydrogenases. *Journal of Biotechnology* 78, 93-113.

[42] Henriksson, G., Zhang, L., Li, J., Ljungquist, P., Reitberger, T., Pettersson, G., and Johansson, G. (2000) Is cellobiose dehydrogenase from *Phanerochaete chrysosporium* a lignin degrading enzyme? *Biochimica et Biophysica Acta - Protein Structure and Molecular Enzymology* 1480, 83-91.

[43] Kumar, R., Singh, S., and Singh, O. V. (2008) Bioconversion of lignocellulosic biomass: Biochemical and molecular perspectives. *Journal of Industrial Microbiology and Biotechnology* 35, 377-391.

[44] Blanchette, R. A., Krueger, E. W., Haight, J. E., Akhtar, M., and Akin, D. E. (1997) Cell wall alterations in loblolly pine wood decayed by the white-rot fungus, *Ceriporiopsis subvermispora*. *Journal of Biotechnology* 53, 203-213.

[45] Bak, J. S., Ko, J. K., Choi, I. G., Park, Y. C., Seo, J. H., and Kim, K. H. (2009) Fungal pretreatment of lignocellulose by *Phanerochaete chrysosporium* to produce ethanol from rice straw. *Biotechnology and Bioengineering* 104, 471-482.

[46] Hirai, H., Sugiura, M., Kawai, S., and Nishida, T. (2005) Characteristics of novel lignin peroxidases produced by white-rot fungus *Phanerochaete sordida* YK-624. *FEMS Microbiology Letters* 246, 19-24.

[47] Kersten, P., and Cullen, D. (2007) Extracellular oxidative systems of the lignin-degrading Basidiomycete *Phanerochaete chrysosporium*. *Fungal Genetics and Biology* 44, 77-87.

[48] Singh, D., and Chen, S. (2008) The white-rot fungus *Phanerochaete chrysosporium*: Conditions for the production of lignin-degrading enzymes. *Applied Microbiology and Biotechnology* 81, 399-417.

[49] Takano, M., Nakamura, M., and Yamaguchi, M. (2010) Glyoxal oxidase supplies hydrogen peroxide at hyphal tips and on hyphal wall to manganese peroxidase of white-rot fungus *Phanerochaete crassa* WD1694. *Journal of Wood Science* 56, 307-313.

[50] Wymelenberg, A. V., Gaskell, J., Mozuch, M., Sabat, G., Ralph, J., Skyba, O., Mansfield, S. D., Blanchette, R. A., Martinez, D., Grigoriev, I., Kersten, P. J., and Cullen, D. (2010) Comparative transcriptome and secretome analysis of wood decay fungi *Postia placenta* and *Phanerochaete chrysosporium*. *Applied and Environmental Microbiology* 76, 3599-3610.

[51] Cameron, M. D., and Aust, S. D. (2001) Cellobiose dehydrogenase - An extracellular fungal flavocytochrome. *Enzyme and Microbial Technology* 28, 129-138.

[52] Wood, J. D., and Wood, P. M. (1992) Evidence that cellobiose:quinone oxidoreductase from *Phanerochaete chrysosporium* is a breakdown product of cellobiose oxidase. *Biochimica et Biophysica Acta - Protein Structure and Molecular Enzymology* 1119, 90-96.

[53] Daniel, G., Volc, J., and Kubatova, E. (1994) Pyranose oxidase, a major source of H_2O_2 during wood degradation by *Phanerochaete chrysosporium*, *Trametes versicolor*, and *Oudemansiella mucida*. *Applied and Environmental Microbiology* 60, 2524-2532.

[54] De Koker, T. H., Mozuch, M. D., Cullen, D., Gaskell, J., and Kersten, P. J. (2004) Isolation and purification of pyranose 2-oxidase from *Phanerochaete chrysosporium* and characterization of gene structure and regulation. *Applied and Environmental Microbiology* 70, 5794-5800.

[55] Morel, M., Ngadin, A. A., Jacquot, J. P., and Gelhaye, E. (2009) Chapter 6 Reactive oxygen species in *Phanerochaete chrysosporium* relationship between extracellular oxidative and intracellular antioxidant systems. pp. 153-186.

[56] Otsuka, Y., Sonoki, T., Ikeda, S., Kajita, S., Nakamura, M., and Katayama, Y. (2003) Detection and characterization of a novel extracellular fungal enzyme that catalyzes the specific and hydrolytic cleavage of lignin guaiacylglycerol β-aryl ether linkages. *European Journal of Biochemistry* 270, 2353-2362.

[57] Masai, E., Ichimura, A., Sato, Y., Miyauchi, K., Katayama, Y., and Fukuda, M. (2003) Roles of the enantioselective glutathione S-transferases in cleavage of β-aryl ether. *Journal of Bacteriology* 185, 1768-1775.

[58] Tanamura, K., Abe, T., Kamimura, N., Kasai, D., Hishiyama, S., Otsuka, Y., Nakamura, M., Kajita, S., Katayama, Y., Fukuda, M., and Masai, E. (2011) Characterization of the third glutathione S-transferase gene involved in enantioselective cleavage of the β-aryl ether by *Sphingobium* sp. strain SYK-6. *Bioscience, Biotechnology and Biochemistry* 75, 2404-2407.

In: Bioenergy Systems, Biological Sources … ISBN: 978-1-62417-330-1
Editor: Michel C. Allard

Chapter 4

BIOMASS AVAILABILITY AND USES IN BRAZIL

***Aline Machado de Castro*[1], *Guilherme Eugênio Machado Lopes*[2], *José Nicomedes Junior*[1] *and Silvia Machado de Castro*[3]**

[1]Biotechnology Department, R&D Center, PETROBRAS, Rio de Janeiro, Brazil

[2]Agricultural Technology Department, PETROBRAS Biofuel, Rio de Janeiro, Brazil

[3]Institute of Geosciences, Federal University of Rio de Janeiro, Rio de Janeiro, Brazil

ABSTRACT

Brazil is in the list of the largest biofuels producers in the world. The country presents biodiversity and environmental characteristics that make it suitable for the cultivation of several biomasses with high productivity, such as sugarcane (*Saccharum* spp.) and soybean (*Glycine max*), which are used for the production of bioethanol and biodiesel, respectively. This chapter discusses cultivation aspects, geographical distribution, production costs and the historical series of the production of the main biomasses cultivated in Brazil, as well as comments on the other agricultural resources that are being studied for the production of biofuels and green chemicals, such as castor fruit (*Ricinus communis*) and oil palm fruit (*Elaeis guineensis*). The current and potential uses of these materials in the context of sustainable products and biorefineries are also discussed.

LIST OF STATE NOMENCLATURE

AC: Acre
AL: Alagoas
AM: Amazonas
AP: Amapá

BA: Bahia
CE: Ceará
DF: Distrito Federal (capital)
ES: Espírito Santo
GO: Goiás
MA: Maranhão
MG: Minas Gerais
MT: Mato Grosso
MS: Mato Grosso do Sul
PA: Pará
PE: Pernambuco
PB: Paraíba
PI: Piauí
PR: Paraná
RJ: Rio de Janeiro
RR: Roraima
RO: Rondônia
RN: Rio Grande do Norte
SC: Santa Catarina
SE: Sergipe
SP: São Paulo
RS: Rio Grande do Sul
TO: Tocantins

1. Introduction

In 1850, biomass represented 85% of the worldwide consumption, but the period between the 19th and 20th centuries was marked with deep changes in the world energy use. With the introduction of steam power machines and internal combustion engines (Otto and diesel cycles) during the industrial revolution, the use of fossil fuels started increase rapidly (IEA Bioenergy, 2010).

Since the last decades of 20th century, however, due to its characteristics of clean and sustainable energy, biomass has been considered a highly viable alternative to diversify sources of energy worldwide. It is estimated that currently, biomass accounts for about 14% of the global consumption of primary energy, which is higher than the participation of coal and similar to those from natural gas and electricity (ANEEL, 2009). Biomass is also responsible for about 75% of all renewable energy available in the world (Kampman et al., 2010) and it is estimated that by the end of this century, it can reach up to 20% of total global consumption (Baitelo, 2007).

In the industrialized countries, biomass is used mainly for electricity and heat, although there is a clear growing use for biofuels production. In the developing countries, biomass accounts for about 22% of the total energy consumption, with peaks of 90% in some countries (IEA Bioenergy, 2012). In Brazil, specifically, the primary renewable sources of energy are biomass (31.3%), hydropower (14.1%) and firewood and charcoal (10%) (Sallet and Alvim, 2011).

In Brazil, the production of biofuels is growing. It is estimated that alcohol fuel (mainly produced from sugarcane) will account for 60% of the transportation fuels in light vehicles in Brazil by 2030 (MME, 2007). It is important to understand, therefore, how the huge amount of agricultural biomass, which includes energy crops, by-products, agricultural and agroindustrial wastes that are generated annually in Brazil can be used.

2. Main Biomasses Produced in Brazil

Brazil is a tropical country with continental dimensions, located in the Southern hemisphere, in Latin America. It is composed of 26 states (including 1 Federal District) that are distributed in five main regions: North, Northeast, South, Southeast and Central-West (Figure 1). Its general physical characteristics (such as climate and topological) are proper for the cultivation of several agricultural materials, thus giving it a unique biodiversity in the world (Carvalho, 2010). Among these biomasses, sugarcane (*Saccharum* spp.) and soybean (*Glycine max*) are some of the most important for the national economy, representing the main raw materials for the production of bioethanol and biodiesel, respectively (Kohlhepp, 2010).

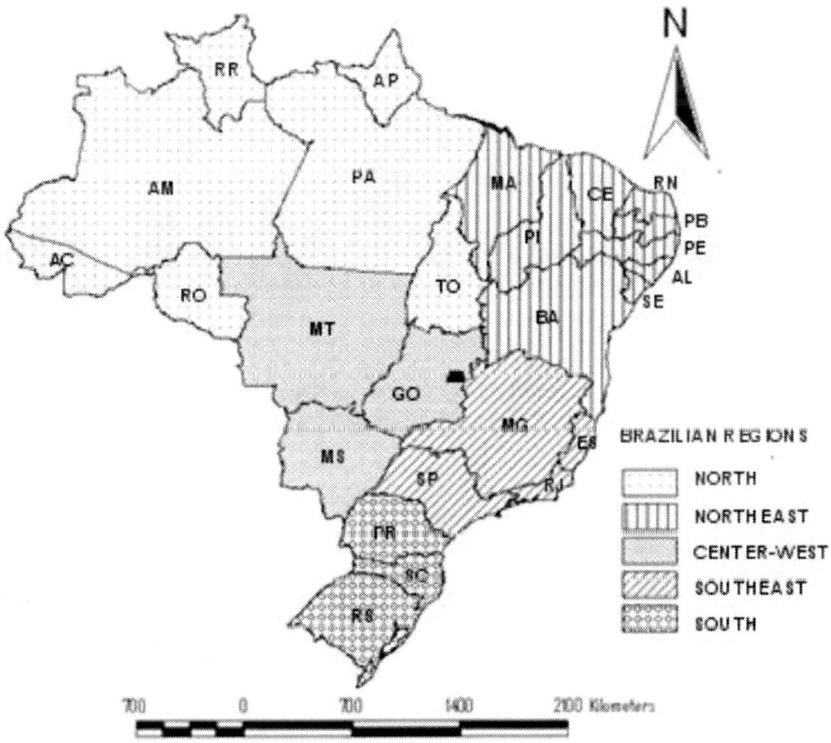

Figure 1. Map of Brazilian states and regions. For abbreviations, see list of abbreviations at the beginning of this chapter.

Brazilian agriculture has been showing in the last decades potential do grow even more when facing demands from a continuously growing population. Innovations in the fields of soil and water use and handling, genetic improvement of plants, biological control of pests, biological safety, as well as the development of processes and practices prove the scientific advance was significant. In addition, the possibility of expansion of area and productivity of several cultivars ensures to Brazil a top position in the ranking of countries with a biomass-based economy. In this section some of the most significant biomasses cultivated in Brazil will be discussed in more details.

Sugarcane

Sugarcane can be considered the main biomass used for energy in Brazil. It is the basis for the agribusiness for ethanol and sugar production, which is represented by more than 350 mills and more than 1 million direct and indirect employments in the whole country (Unica, 2012).

Sugarcane cultivation is secular in Brazil, but has been mainly encouraged at a commercial level since the 1970s. In 1975 the Brazilian government created the National Alcohol Program (named Pro-álcool), which afforded the development of a strong industry based in the use of sugarcane for ethanol production, by means of research and development in the fields of agrosciences, bioprocess technology, as well as mechanical engineering for the development of engines to use the hydrous alcohol (96% v/v ethanol content) as a fuel. Pro-álcool can be considered one of the biggest programs in the world for commercial use of biomass for energy (Nitsch, 1991). In 2011, Brazil was estimated to have about 14 million new vehicles based on flex-fuel technology, which can use ethanol as fuel, representing 46% of the total light vehicles in the country (Brasil, 2012).

One of the most important factors that contributed for the success of the sugarcane-based industry in Brazil, in addition to the investments in research and development, was the structure of the productive sector, allowing dissemination of the knowledge. The excellence of Brazilian programs for the genetic improvement was a key-point to reach the bioeconomy that Brazil presents today. Currently, sugarcane varieties that are cultivated in large scale are result of the interspecific crossing between the species *Saccharum officinarum, S. barbieri, S. sinense* and the wild species *S. spontaneum* and *S. robustum*. Such varieties present some of the most complex genomes in the plant kingdom (D'Hont et al., 1996). This complexity is a barrier for the use of conventional techniques for the genetic improvement of the culture (Hogarth, 1987), thus making sugarcane an excellent candidate for the improvement by means of genetic engineering.

The use of genetic engineering strategies for the improvement of sugarcane varieties has been growing in the last years. One example is the program SucEST (Sugarcane Expressed Sequence Tags), funded by the São Paulo (SP) state government. The program afforded the largest and most complete sequence of expressed genes in sugarcane ever done. SucEST researches resulted in high-level scientific data, which opened new frontiers to unravel molecular mechanisms and the biochemistry of sugarcane in order to understand gene functions responsible for the superiority characteristics aimed by the researchers (Calsa Junior et al., 2004).

Oleaginous Plants

In addition to sugarcane, some oleaginous plants are also significant for the biomass-based economy, since vegetable oils are the main feedstocks used for biodiesel production in Brazil. The Brazilian government launched in 2004 the National Programm for Production and Use of Biodiesel (PNPB), driving investments for the implementation of a sustainable program based on social inclusion, guarantee of competitive prices, quality and supply in the biodiesel production chain from diverse crops throughout the country (PNPB, 2010).

There are hundreds of oleaginous plants available for cultivation in Brazil. However, the production for commercial use as food has been focused on soybean (*Glycine max*), cotton (*Gossypium hirsutum*), peanuts (*Arachis hypogaea*), sunflower (*Helianthus annus*), canola (Brassica napus), sesame (*Sesamum indicum*), corn (*Zea mays*), olive (*Olea europaea*), oil palm (*Elaeis guineensis*) and coconut (*Cocos nucifera*) (Trzeciak et al., 2008). The cultivation of feedstocks for biodiesel production presents potential in terms of the creation of employments and social inclusion, especially for local farmers. In the Brazilian semiarid region (mainly Northeast region) and in the North region, population is even eager to be socially included, and this can be reached by means of the cultivation of oleaginous plants such as castor (*Ricinus communis* L.) and jatropha (*Jatropha curcas*).

In general terms, what is observed is the concentration of soybean cultivation in South, Southeast and Center-west regions, whereas castor is more cultivated in the Northeast and oil palm in the North. Other oleaginous, such as sunflower, peanuts and cotton present some participation, but in a lower extent and in smaller regions (IBGE, 2012).

Brazil is one of the largest soybean producers in the world. Soybean is the main source of vegetable oil for the production of biodiesel, accounting for about 80% of all biodiesel produced in the country (Sallet and Alvim, 2011). The use of soybean for energetic purposes had raised discussions concerning impacts on food resources. However, in Brazil it was reached equilibrium for its use for both industries.

Castor is another oleaginous crop considered for biodiesel production in Brazil. Worldwide, it is not a relevant feedstock for this purpose, being responsible for less than 0.5% of the global market of vegetable oils (Furlan Junior, et al. 2006). Although Brazil is the third largest castor oil producer in the globe, production and productivity in Brazil have remained below expectations despite the growth of both domestic and global demands for the castor oil, which is highly valued due to its many industrial uses.

One of the main characteristics of castor plant is its resistance to lack of water, thus allowing it to be cultivated in an economical fashion in the Brazilian semiarid region, where there are few alternative biomasses (Vaz et al., 2010).

It is a consensus that in order to achieve global production levels of castor fruits, it will be necessary to have mutual efforts from different players. The scientific community needs to work in a collaborative manner to match the bottlenecks regarding production, processing and commercialization of castor oil, which include: i. Characterization and exchange of genetic materials; ii. Studies on heritability of some important agronomic characteristics; iii. Development of technologies for mechanization of castor fruit harvesting; iv. Development of strategies for an integrated control of pests, weeds and diseases; v. improvement of tools for the determination of ricin toxicity; and vi. Development of a global model comprising technical, economic and market variables that impact castor fruit production (Severino, 2012).

Oil palm is another potential feedstock for biodiesel production in Brazil, mainly in the North region. Its large scale cultivation started during the 1960s in Malaysia and is currently expanding in the world, being the Amazonas state (AM) one of the most promising places in Brazil for its cultivation due to climate and soil characteristics. The cultivation of oil palm in this state can also present social benefits, since local producers can explore the tree with sustainability and minimum impact to the environment (Nunes and Alves, 2010).

Oil palm tree produces oil with outstanding quality, but there is a bottleneck regarding integration between the harvesting areas and industrial processing units, since the oils tends to be rapidly acidified, thus not qualified for chemical transesterification reactions (Curvelo, 2010). Other bottlenecks, however, need also to be addressed: i. Low capacity to produce seeds in Brazil; ii. Low investments in research and development; and iii. Low availability of specialized human resources (Furlan Junior et al., 2006).

More details regarding the production of the abovementioned biomasses in Brazil will be discussed in the next section.

3. Production Statistics

The production of the selected biomasses discussed in the previous section in Brazil was obtained from searches in the Brazilian Institute of Geography and Statistics (IBGE) database, which is supported by the Brazilian Government. Figure 2 shows the historical series of the production of sugarcane bagasse, castor fruit (which contains the seeds), oil palm fruit (which comprises the seeds and pulp) and soybean grains in the last two decades. The difference in the magnitudes can be clearly observed: while sugarcane bagasse and soybean are produced annually in the order of dozens to hundreds of million tons, oil palm fruit and castor fruit varied in the range of thousand tons. With the exception of castor fruit, which production fluctuated over the seasons, the production of the other three feedstocks evaluated was continuously increasing over the years (IBGE, 2012).

When compared to the worldwide production, the Brazilian participation in the sugarcane production increased from 24.9% in 1990 (1.05 trillion tons in the world) to 42.6% in 2010 (1.69 trillion tons globally). Regarding soybean, Brazil was responsible for 18.3% of all grains produced in the world in 1990 (108 million tons) and in 2010, the participation had been increased to 26.3% (261 million tons in the world) (Faostat, 2012; IBGE, 2012).

The yield has also been increasing over the last two decades (Figure 3). In 1990, the yield for soybean grains was 1731 kg/ha and in 2010 it reached 2947 kg/ha, an increase of 70%. Regarding sugarcane, the yield was increased from 61.5 t/ha in 1990 to 79.0 t/ha in 2010 (28.6% increase). In both cases, the improvement can be attributed to investments in new machinery for harvesting as well as genetic manipulation of the varieties. In this sense, it should be also informed that the number of cuts from a same sugarcane seedling planted increased from 3 to 6 in the last 30 years, thus giving benefits in the process yield (Neves, 2011). It can be also observed from Figure 3 an average yield improvement for oil palm fruits of 65% from 1990 to 2010, indicating the potenciality of this oleaginous plant, mainly as compared to soybean yields. The yields for castor fruit are still low in Brazil (ranged from 0.26-0.77t/ha in the period evaluated), however, as discussed previously, the culture presents a great potentiality for local farmers in the semiarid region.

Production prices are also quite different between the biomasses evaluated (Figure 4). Sugarcane prices have increased from 10.9 BRL/t to 39.46 BRL/t (equivalent to about 19 USD/t) between 1994 and 2010. Soybean grains prices raised, in the same period, from 154 BRL/t to 544 BRL/ton, whereas castor bean fruit increased from 181BRL/t to 953 BRL/t (24.2% increase only from 2009 to 2010). Oil palm fruit prices increased 83% in the same period (from 98 BRL/t to 180 BRL/t) (IBGE, 2012). It should be informed that for this indicator, the period investigated was from 1994 and not 1990, in order to make comparisons easier, since Brazilian currencies changed in 1994.

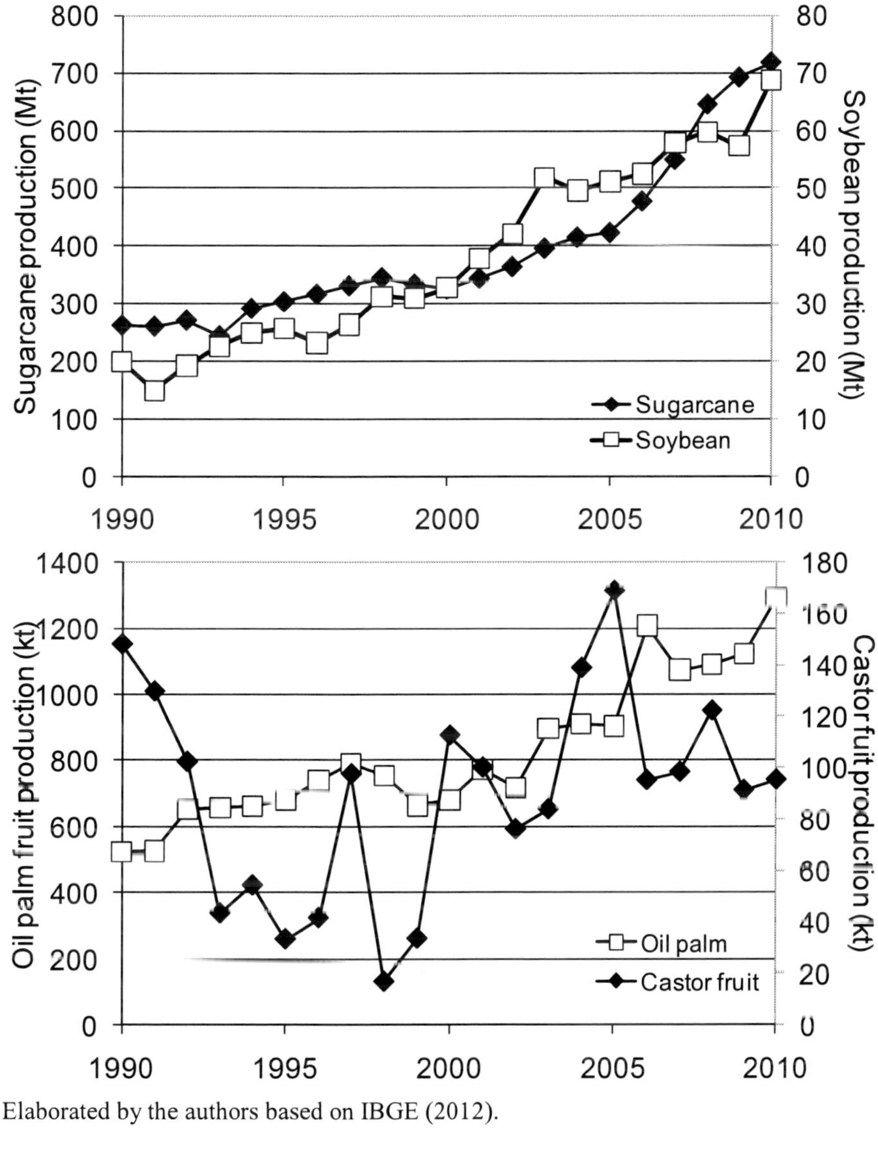

Source: Elaborated by the authors based on IBGE (2012).

Figure 2.Historical series for the production of sugarcane, castor fruit, oil palm fuit and soybean in Brazil.

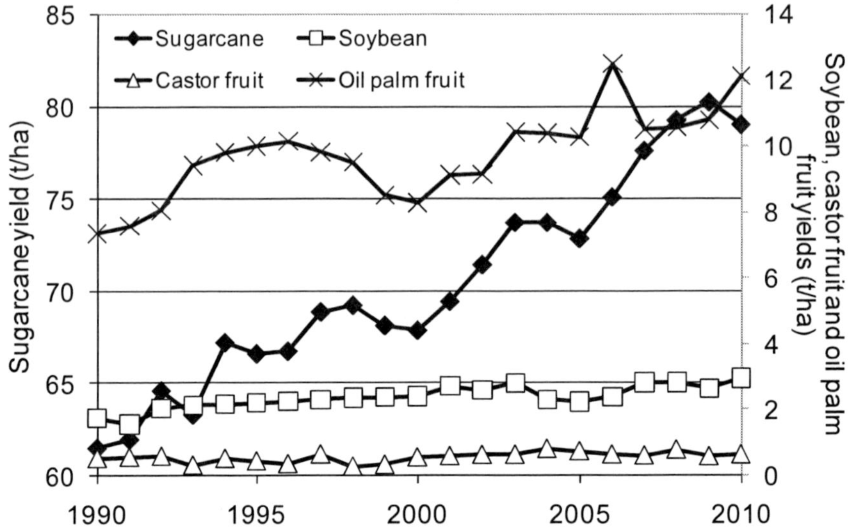

Source: Elaborated by the authors based on IBGE (2012).

Figure 3. Historical series for the yields of sugarcane, castor fruit, oil palm fruit and soybean in Brazil.

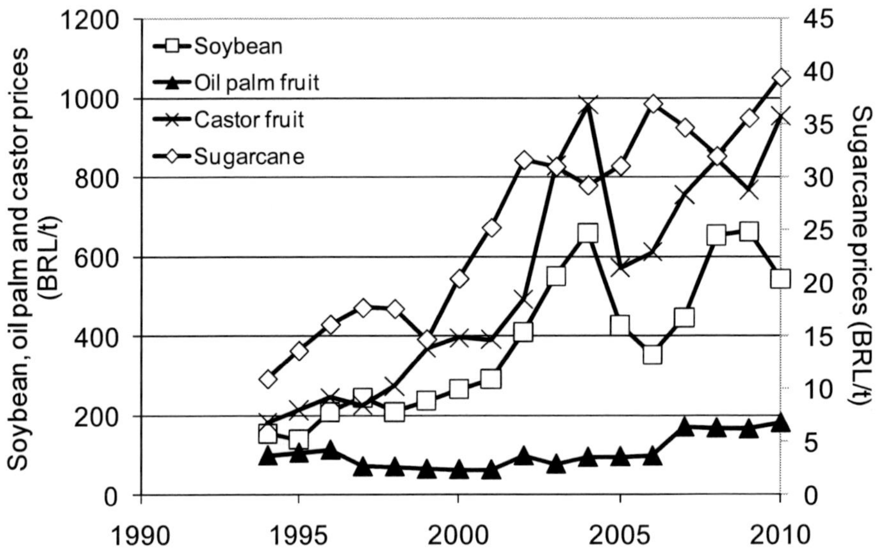

Source: Elaborated by the authors based on IBGE (2012).

Figure 4. Historical series for the prices of sugarcane, castor fruit, oil palm fruit and soybean in Brazil.

Sugarcane production is more concentrated in the Southeast region (mainly São Paulo state), which comprised in 2010 69.5% of all sugarcane harvested in Brazil. As can be seen in Figure 5 the participation of the Southeast and Center-west regions have been increasing over the last two decades, while in the Northeast region the percent sugarcane production has been decreasing (from 27.3% to 9.6% between 1990 and 2010). The production of sugarcane in the North region was always below 1% in the period investigated.

For the season 2012/2013, the Sugarcane Industry Association estimates a production of sugarcane in the Southeast/South regions of 518 million tons, corresponding to approximately 21.05 billion liters of ethanol produced (Unica, 2012).

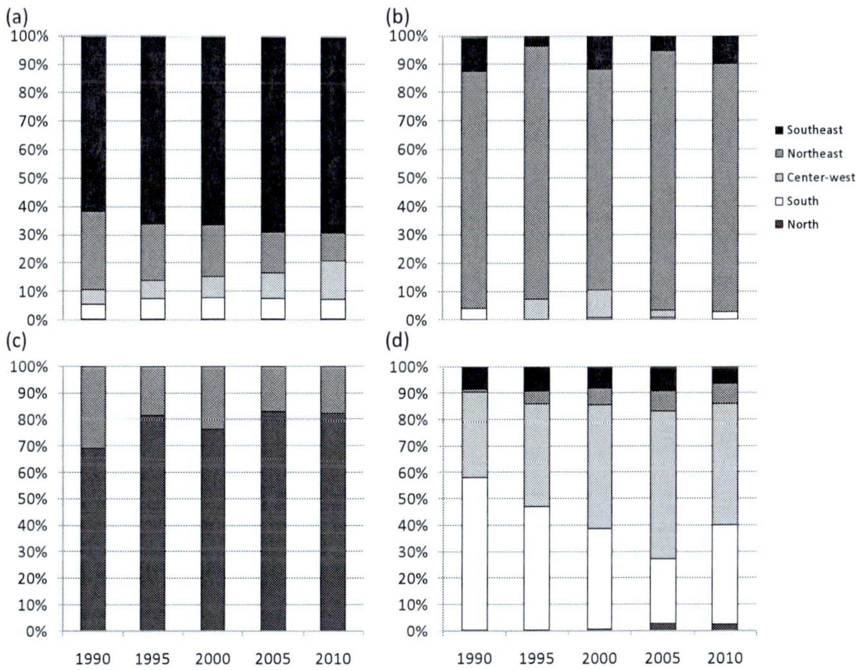

Source: Elaborated by the authors based on IBGE (2012).

Figure 5. Distribution of biomasses production in Brazil, by region: (a) sugarcane; (b) castor seed; (c) oil palm; and (d) soybean. For Brazilian production quantities per year, please see Figure 2.

Brazil has already been the major castor fruit producer in the world, in 1978/1982, when it was responsible for 32% of the fruits harvested. The biomass was not properly encouraged since then and in 2004 and 2005 seasons the country accounted for only 11% and 13% of the global production, respectively (Santos and Kouri, 2006). In Brazil, 78% of the production in 2010 occurred in Bahia (BA) state, in the Northeast region, where at least 77.7% (observed in 2000) of the castor fruits harvested in Brazil between 1990 and 2010 were obtained. North region did not produce castor fruits over the last two decades (IBGE, 2012).

Oil palm fruit production was observed only in the Northeast and North regions from 1990 to 2010 (Figure 3c). The total Brazilian production is less than 1% of the world production, which rose from 60.9 million tons to 210.9 million tons in the same period abovementioned, with a significant production in Malaysia (Faostat, 2012). It is estimated that Brazil produced in 2005 only 170 thousand tons of palm oil, which is less than 0.1% of the area with potential for the cultivation of this biomass.

Regarding soybean production by region (Figure 3d) it is interesting to observe that there was a change throughout the last two decades. In 1990, South was the main producing region, with very low production in the Northeast. Since 2000, Center-west has been the major Brazilian region producing the grains and the participation of the North region has increased from 0.2% to 2.4% between 1990 and 2010.

4. Current and Potential Uses of Biomass and its Co-Products

As shown in the previous section, some biomasses are produced in very high amounts in Brazil. The harvesting and industrial processing operations applied on them can generate byproducts, co-products and residues, which can be used in further processes. In this section some examples will be presented.

Considering the production of sugarcane in Brazil in 2010 (717 million tons), it can be inferred that large amounts of sugarcane straw and bagasse are generated. Castro and Castro (2012) recently reviewed that almost 100 million tons of sugarcane bagasse (approximately 50% moisture) were produced in Brazil in 2008 due to sugarcane stalks processing (crushing). Both sugarcane bagasse and sugarcane straw present lignocellulosic composition, being cellulose, hemicelluloses and lignin their major fractions (Rocha et al., 2011; Pandey et al., 2000).

Such materials can be used in a vast number of applications, via biochemical (e.g. hydrolysis, fermentation) or thermochemical (e.g. gasification, pyrolysis) processes, or even the combination of both (Clements and Van Dyne, 2006). By using enzymes, the polysaccharidic fractions of lignocellulose (cellulose and hemicelluloses) are converted to free sugars that can be subsequently fermented to a large variety of bioproducts, including solvents, organic acids and alcohols, as shown in Table 1.

Pyrolysis of sugarcane bagasse can also be combined with biotechnological conversion. As reviewed by Brown (2006), the bio-oil obtained via fast pyrolysis process presents levoglucosan, which can be hydrolysed to free sugars and then fermented to ethanol. Other thermochemical process is the gasification, where a synthesis gas (comprised mainly of CO, H_2, CO_2) can be fermented to organic acids (butyric and acetic), alcohols (ethanol and butanol) and polyesters (Brown, 2006).

The combination of biotechnological and chemical steps can also be used for the production of valuable bioproducts. As described by Girisuta (2007) glucose released by enzymatic hydrolysis of cellulose can be chemically converted to levulinic acid, via dehydration of hydroxymethylfurfural.

Other uses for lignocellulosic materials are as animal feed, as a substrate for composting, in the manufacture of particle boards, as organic component for soil, for the production of furfural and other chemicals (Couto et al. 2004), in the formulation of composites (Carvalho Neto et al., 2009) and for the production of biodegradable polymers as polyhydroxyalkanoates (Satyanarayana et al., 2009). A mixture of crushed sugarcane bagasse and sugarcane filtercake has been used successfully to produce seedlings of different plant species (Serrano et al., 2006). Sugarcane bagasse can also be used for the production of enzymes (cellulases), which present numerous industrial applications (Castro et al., 2010).

Currently the primary use of sugarcane bagasse in Brazil is as solid fuel for thermal energy generation in sugarcane mills in order to co-generate electricity (ANEEL, 2009). There is, though, a dilemma about the best use for the bagasse, since its use for the production of second generation bioethanol is prominent in the biofuels industry.

The production of second generation ethanol from various biomasses, including sugarcane bagasse, has required much efforts and investment in research and development.

According to Rosa and Garcia (2012) several methods have been developed with that purpose.

Table 1. Potential uses of sugarcane bagasse and other lignocellulosic biomasses via fermentation

Feedstock	Products	Reference
Corn cob	Acetone, n-butanol, ethanol (16g/L total)	Zhang et al. (2012)
Rice bran	Acetone, n-butanol, ethanol (12g/L total)	Al-Shorgani et al. (2012)
Soybean fiber and DDGS	Animal feed	Lio and Wang (2012)
Rice straw	2,3-butanediol (24.6g/L)	Wong et al. (2012)
Sweet sorghum bagasse	Ethanol (6.12g/L)	Cao et al. (2012)
Sugarcane bagasse	Ethanol (14g/L, from cellulose) Ethanol (19g/L, from hemicellulose)	Zhu et al. (2012) Cheng et al. (2008)
Sugarcane bagasse	Xylitol (31.9g/L, from hemicellulose)	Roberto et al. (1991)
Sugarcane bagasse	Succinic acid (22.5g/L, from hemicellulose)	Borges and Pereira Jr (2011)
Sugarcane bagasse	Lactic acid (10.8g/L, from cellulose)	Laopaiboon et al. (2010)
Sugarcane bagasse	Acetone, n-butanol, ethanol (9.44g/L, total)	Lu (2011)
Sugarcane bagasse	Citric acid (12.8g/L, from cellulose)	Al-Mahin et al. (2012)
Sugarcane bagasse	Lipids (6.68g/L, from cellulose and hemicellulose)	Tsigie et al. (2011)

Table 2. Protein content of some biomasses

Biomass	Protein content (%)	Reference
Sunflower bran	33.4	Mendes et al. (2005)
Wheat bran	17.7	Mathlouthi et al. (2002)
Babassu flour	2.3	Baruque Filho et al. (2000)
Peanut meal	59	Balan et al. (2009)
Canola meal	49.8	Balan et al. (2009)
Sesame meal	41.5	Balan et al. (2009)
Castor seed cake	28.7-42.5	Freire and Nóbrega (2006); Melo et al. (2008); Severino (2005)
Jatropha cake	28.7	Souza et al. (2009)
Oil palm cake	14.5-19.6	Alimon (2004)
Soybean meal	51.0	Balan et al. (2009)

However, in order to use efficiently the sugarcane straw, more studies are needed to better understand the phenomena involved in the biochemical conversion and some challenges still need to be overcome, such as the definition of the conditions for biomass pretreatment and hydrolysis, to finally release fermentable sugars.

As reviewed by Peng et al. (2012), hemicelluloses, the second major sugars sources in sugarcane bagasse, are heteropolysaccharides which may comprise sugars as glucose, xylose, arabinose, galactose, rhamnose and mannose, as well as uronic acids (galacturonic and glucuronic). This fraction of lignocellulosic materials can also provide, therefore, a vast portfolio of substrates that can be converted by fermentative processes to ethanol and organic acids, amongst other molecules, as shown in Table 1.

Regarding the oleaginous biomasses, soybean is the crop of greatest economic impact in Brazil. In the cultivation area only the grains are harvested, and their processing result in a protein-rich meal (actually, the main product of the chain), which can be used in animal feed, as well as oil as a byproduct, used in the food industry (Balan et al., 2009), and most recently as the main feedstock for the biodiesel industry. Table 2 summarizes the protein content of several biomasses, proving that soybean bran presents a very interesting nutritional content for use as animal feed.

During the harvesting in the field some biomasses are left in the soil, in a proportion of about 2.5 tons per ton of soybean harvested. These include straw and stalks (mostly lignocellulosic) and were estimate to be in the order of 47.8 million tons (Mamedes et al., 2010). However, Brazilian soybean is almost entirely grown in no-tillage system, so it is important that part of the crop residues remain in the soil surface to ensure the sustainability of the system, so that the removal of this residual biomass from the field may not be interesting agronomically and environmentally. Moreover, the lack of mechanical harvesting technology, that does not allow recovery of the straw, as well as the high costs that would be associated with the removal and transportation of the lignocellulosic residues does not seem to be currently economically viable.

Not so disseminated as soybean in Brazil, but also with a huge potential, oil palm is a culture on the rise in the country that present a growth expectation for the coming years. In addition to crude palm oil (20%) and palm kernel oil (5%), during the industrial processing of palm fruits are also generated stems (3.5%), fibers (22%), husks (12%) and palm kernel meal (1.5%) (Furlan Junior, 2006). Similarly to soybean harvesting, during the recovery of oil palm fruits the leaves of the palm tree remain in the field, as well as the residual lignocellulosic parts obtained during the process of pruning and removal of old tree trunks (stems). It is estimated that approximately 184 million tons of residual biomass are generated worldwide due the cultivation of oil palm every year (Mohammed et al., 2011).

The lignocellulosic parts of the oil palm tree can be used for ethanol production via enzymatic hydrolysis and fermentation, and the alcohol can be used for transesterification reaction of the palm oil, in an integrated fashion, as proposed by Gutiérrez et al. (2009) and more recently reported by Piarpuzán et al. (2011).

The current employment opportunities regarding the production chain of oil palm include the use of stalks and fibers in boilers, the return to the field for use as mulch, compost and fertilizer. The shell, with high calorific value, can be used as fuel or as raw material for coal plants. Palm kernel cake (meal), with a protein content of 15%, can be destined for animal feed. However, due to variability in biomass and certain intrinsic characteristics such as high moisture content, low energy density, hygroscopic behavior, difficulty of storage, there are still some challenges to overcome for its use as an energy resource (Aziz et al., 2012).

Finally, concerning castor fruit, the main byproduct of the production chain of its oil is the cake, which main application is currently as an organic fertilizer. Castor seed cake presents high levels of nitrogen, fast mineralization and nematicide properties (Severino et al., 2012). Despite presenting a high potential for use in animal feed, the use of cake in this area is still prohibitive because it presents toxic and allergenic compounds such as the protein ricin, thus requiring detoxification prior to its use. Several methods of cake detoxification, among physical, chemical and biological processes, have been investigated with promising results, at least on a small scale. However, no process has yet been successfully applied in an industrial level, due to costs and the lack of fast and reliable methods for monitoring the detoxified material (Severino et al., 2012). Other research areas that aimed at the valorization of the castor seed cake includes lipase production (Godoy et al. 2011), simultaneous amylase, protease, cellulase and xylanase production (Castro et al., 2011) and ethanol production by the use of acid or enzymatic hydrolysis (Melo et al. 2008).

The castor fruit husk, also called capsule, is a material that is currently discarded or used as fertilizer or fuel for heat generation. For each ton of castor fruit processed, it is estimated that are generated around 620 kg of husk (Severino et al., 2005). It is estimated that in 2005 were generated in Brazil, approximately 130,000 tons of castor fruit husks, from the castor oil production (Lima et al., 2008) and in 2010 a global production around 830,000 tons (Severino et al., 2012). Studies on physical and chemical characterization of this material revealed that when fresh, it presents high levels of electrical conductivity and possibly the presence of phytotoxic phenolic compounds. However, after composted it has physical and chemical properties suitable for use as a substrate for plants, either pure or in a mixture for the production of seedlings from seeds and cuttings (Lopes et al., 2012). Other potential uses of the castor fruit husk include animal feed (Santos et al., 2011) and briquettes (Peres et al., 2004).

The use of agricultural and agroindustrial residues as substrates and organic fertilizers contributes for the reduction of production costs and the use of waste whose disposal into the environment represents a strong negative impact. However, their use as a substrate for growing plants may incur in phytotoxicity, salinity change and nutritional or metabolic disorders (Ortega et al. 1996). Some of these problems can be minimized or eliminated by the use of composting techniques (Yates and Rogers, 1981) or by simply washing the material (Carrijo et al. 2002).

The potential applications of biomasses from sugarcane and oleaginous plants processing cited in this section meet a growing demand for more sustainable processes, in all the three pillars reported by Kamm et al. (2006): biobased products, bioenergy and biofuels.

Conclusion

There are several opportunities for the development of a strong biorefinery industry in Brazil. Some biomasses, as sugarcane, soybean, oil palm and castor fruit were shown in this chapter to have potential, due to the large amount they are generated in Brazil, as well as their proper composition to be used in diverse applications, including as fertilizer, animal feed and as raw materials for a sugar-based biorefinery. The learning curve over the last decades in Brazil has shown that it is worthy to invest in research and development, and these investments must continue.

References

Alimon, A.R. The nutritive value of palm kernel cake for animal feed. 2004. Available from: http://www.chgs.com.my/chinese/download/Palm%20Oil%20Developments/june%202004/The%20Nutritive%20Value%20of%20Palm%20Kernel%20Cake%20for%20Animal%20Feed.pdf.

Al-Mahin, A.; Sharifuzzaman, A.B.M.; Faruk, M.O.; Kader, M.A.; Alam. J.; Begum, R.; Or-Rashid, H. Improved citric acid production by radiation mutant *Aspergillus niger* using sugarcane bagasse extract. *Biotechnology*. 2012, 11(1), 44-49.

Al-Shorgani, NKN; Kalil, MS; Yusoff, WMW. Biobutanol production from rice bran and de-oiled rice bran by *Clostridium saccharoperbutyl-acetonicum* N1-4. *Bioprocess and Biosystems Engineering,* 2012, 35(5), 817-826.

ANEEL. *Atlas da energia elétrica do Brasil.* 3ª Ed. 2009. Available from: www.aneel.gov.br.

Aziz, MA; SabiL, KM; Umura, Y; Ismail L. A study on torrefaction of oil palm biomass. *Journal of Applied Sciences*. 2012, 12, 1130-1135.

Baitelo, R. [R]evolução energética: Perspectiva para uma energia global sustentável. *Democracia Viva. Brasil.* 2007. Available from: http://www.greenpeace.org.br/energia /pdf/cenario_brasileiro.pdf.

Balan, V.; Rogers, C.A.; Chundawat, S.P.S. et al. Conversion of extracted oil cake fibers into bioethanol including DDGS, canola, sunflower, sesame, soy and peanut for integrated biodiesel processing. *Journal of the American Oil Chemical Society. 2009*, 86, 157-165.

Baruque Filho, E.A.; Baruque, M.G.A.; Sant'Aanna Jr, G.L. Babassu coconut starch liquefaction: an industrial scale approach to improve conversion yield. *Bioresource Technology*. 2000, 75, 49-55.

Borges, E. R.; Pereira Jr., N. Succinic acid production from sugarcane bagasse hemicellulose hydrolysate by *Actinobacillus succinogenes*. *Journal of Industrial Microbiology and Biotechnology*. 2011, 38, 1001-1011.

Brasil. *Biocombustíveis*. 2012. Available from: http://www.brasil.gov.br/ sobre/economia /energia/matriz-energetica/biocombustiveis/print.

Brown, R.C. Biorefineries based on thermochemical processing. In: Kamm, B.; Gruber, P.R.; Kamm, M. Biorefineries – Industrial Processes and Products. v. 1. *Weinheim: Wiley-VCH Verlag GmbG & Co;* 2006; 227-252.

Calsa Júnior, T.; Carraro, D.M.; Benatti, M.R.; Barbosa, A.C.; Kitajima, J.P.; Carrer, H. Structural features and transcript-editing analysis of sugarcane (*Saccharum officinarum* L.) chloroplast genome. *Current Genetics*. 2004, 46(6), 366-373.

Cao, W; Sun, C; Liu, R; Yin, R.; Wu, X. Comparison of the effects of five pretreatment methods on enhancing the enzymatic digestibility and ethanol production from sweet sorghum bagasse. *Bioresource Technology,* 2012, 111, 215-221.

Carrijo, O.A.; Liz, R.S.; Makishima, N. Fibra da casca do coco verde como substrato agrícola. *Horticultura Brasileira*. 2002, 20(4), 533-535.

Carvalho Neto, A.G.V.; Ganzerlli, T.A.; Fávaro, S.L, Suguino, B.S.; Carvalho, G.M.; Radovanovic, E. *Preparação e caracterização de compósitos de polietileno pós-consumo reforçados com fibras do bagaço de cana.* 10° Congresso Brasileiro de Polímeros. Available from: http://www.ipen.br/ biblioteca/cd/cbpol/2009/PDF/1123.pdf.

Castro, A.M.; Andréa, T.V.; Carvalho, D.F. Teixeira, M.M.P.; Castilho, L.R.; Freire, D.M.G. Valorization of residual agroindustrial cakes by fungal production of multienzyme complexes and their use in cold hydrolysis of raw starch. *Waste and Biomass Valorization*. 2011, 2, 291-302.

Castro, A.M.; Carvalho, M.L.A.; Leite, S.G.F.; Pereira Jr., N. Cellulases from *Penicillium funiculosum*: production, properties and application to cellulose hydrolysis. *Journal of Industrial Microbiology and Biotechnology*. 2010, 37, 151-158.

Castro, S.M.; Castro, A.M. Assessment of the Brazilian potential for the production of enzymes for biofuels from agroindustrial materials. *Biomass Conversion and Biorefinery.* 2012, 2, 87-107.

Cheng, K.K.; Cai, B.Y.; Zhang, J.A.; Ling, H.Z.; Zhou, Y.J.; Ge, J.P.; Xu, J.M. Sugarcane bagasse hemicelluloses hydrolysate for ethanol production by acid recovery method. *Biochemical Engineering Journal.* 2008, 38(1), 105-109.

Clements, L.D.; Van Dyne, D.L. The lignocellulosic biorefinery – A strategy for returning to a sustainable source of fuels and industrial organic chemicals. In: Kamm, B.; Gruber, P.R.; Kamm, M. *Biorefineries – Industrial Processes and Products.* v. 1. Weinheim: Wiley-VCH Verlag GmbG & Co; 2006; 227-252.

Couto, L.C.; Couto, L.; Watzlawick, L.F.; Câmara, D. Vias de valorização energética da biomassa. *Biomassa & Energia.* 2004. 1(1), 71-92.

Curvelo, F.M. Uma imersão no tabuleiro da baiana: o estudo do óleo de palma bruto (*Elaeis guineensis*). Master thesis. *Federal University of Bahia.* 2010.103p.

D'hont, A.; Grivet, L.; Feldmann, P.; Rao, S.; Berding, N.; Glaszmann, J.C. Characterisation of the double genome structure of modern sugarcane cultivars (*Saccharum* spp.) by molecular cytogenetics. *Molecular Genetics and Genomics.* 1996, 250, 405-413.

Faostat. *Production crops.* 2012. Available from: http://faostat.fao.org.

Freire, R.M.M.; Nóbrega, M.B.M. Sistemas de Produção. 2006. Available from: http://sistemasdeproducao.cnptia.embrapa.br/FontesHTML/ Mamona/CultivodaMamona_2ed/torta.html

Furlan Júnior, J. Dendê: manejo e uso dos subprodutos e dos resíduos. *Embrapa Amazônia Oriental.* 2006. 37p.

Furlan Junior, J.; Kaltner F.J.; Azevedo, G.F.P.; Campos, I.A. *Biodiesel: porque tem que ser dendê.* Embrapa Amazônia Oriental. 2006.

Girisuta, B. *Levulinic acid from lignocellulosic biomass. DSc thesis.* 2007. Available from: http://dissertations.ub.rug.nl/FILES/faculties/science/ *2007/b.girisuta/thesis.pdf.*

Godoy, M.G.; Gutarra, M.L.E.; Castro, A.M.; Machado, O.L.T.; Freire, D.M.G. Adding value to a toxic residue from the biodiesel industry: Production of two distinct pool of lipases from *Penicillium simplicissimum* in castor bean waste. *Journal of Industrial Microbiology and Biotechnology.* 2011, 38, 945–953.

Gutiérrez, L.F.; Sánchez, O.J.; Cardona, C.A. Process integration possibilities for biodiesel production from palm oil using ethanol obtained from lignocellulosic residues of oil palm industry. *Bioresource Technology.* 2009, 100, 1227 1237.

Hogarth, D. M. Genetics of sugarcane. In: Heinz, D. J., editor. *Sugarcane Improvement Through Breeding.* New York:Elsevier; 255-272; 1987.

IBGE. *Produção Agrícola Municipal.* 2012. Available from: http://www.ibge.gov.br

IEA Bioenergy. *Bioenergy – a Sustainable and Reliable Energy Source.* 2009 aug. Available from: http://www.ieabioenergy.com/MediaItem.aspx? id=6360.

IEA Bioenergy. IEA Bioenergy Annual Report 2011. 2012 feb. Available from: http://www.ieabioenergy.com/LibItem.aspx?id=7315.

Kamm, B.; Kamm, M.; Gruber, P.R.; Kromus, S. Biorefinery systems – An overview. In: Kamm, B.; Gruber, P.R.; Kamm, M., eds. *Biorefineries – Industrial Processes and Products.* v. 1.Weinheim: Wiley-VCH Verlag GmbG & Co; 2006; 227-252.

Kampman, B.; Bergsman, G.; Schepers, B. et al. (2010) *BUBE: Better Use of Biomass for Energy.* 2010. 151p. Available from: http://www.ce.nl/ publicatie/bube%3A_better_use _of_biomass_for_energy%3Cbr%3Ebackground_report_to_the_position_paper_of_iea_redt_and_iea_bioenergy/1104.

Laopaiboon, P.; thani, A.; Leelavatchamaras, V.; Laopaiboon, L. Acid hydrolysis of sugarcane bagasse for lactic acid production. *Bioresource Technology*. 2010. 101(3), 1036-1043.

Lio, J; Wang, T. Solid-state fermentation of soybean and corn processing coproducts for potential feed improvement. *Journal of Agricultural and Food Chemistry,* 2012, 60(31), 7702-7709.

Lopes, G.E.M.; Vieira, H.D.; Jasmim, J.M.; Shimoya, A.; Marciano, C.R. Casca do fruto da mamoneira como substrato para as plantas. *Revista Ceres*. 2011, 58(3), 350-358.

Lu, C. Butanol production from lignocellulosic feedstocks by acetone-butanol-ethanol fermentation with integrated product recovery. DSc thesis. 2011. Available from: http://etd.ohiolink.edu/view.cgi?acc_num=osu13068 23156.

Mamedes, J.A.; Rodrigues, M.P.J.; Vanissang, C.A. Biomassa no Brasil. *Revista de divulgação do Projeto Universidade Petrobras e IF Fluminense*. 2010, 1, 65-73.

Mathlouthi, N.; Saulnier, L.; Quemener, B. et al., Xylanase, β-glucanase, and other side enzymatic activities have greater effects on the viscosity of several feedstuffs than xylanase and β-glucanase used alone or in combination. *Journal of Agricultural and Food Chemistry*. 2002. 50, 5121-5127.

Melo, W.C.; Santos, A.S.; Anna, L.M.M.; Pereira, N. Acid and enzymatic hydrolysis of the residue from castor (*Ricinus communis* L.) oil extraction for ethanol production: Detoxification and biodiesel process integration. *Journal of the Brazilian Chemical Society*. 2008, 19, 418–425.

MME (Ministério de Minas Energia). *Matriz Energética Nacional* 2030. 2007. 254p. Available from: http://www.mme.gov.br/mme/menu/todas_ publicacoes.html.

Mohammed, MAA; Salmiaton, A; Wan Azlina, WAKG; Mohammad Amran, MS; Fakhru'l-Razi, A; Tufiq-Yap, YH. Hydrogen rich gas from oil palm biomass as a potential source of renewable energy in Malaysia. *Renewable and Sustainable Energy Reviews*. 2011, 15, 1258-1270.

Neves, P.T. Manutenção produtiva total (total productive maintainance): estudo de caso na colheita mecanizada de cana-de-açúcar (*Saccharum* spp.). *MSc dissertation*. 2011. Available from: *www.teses.usp.br/teses*

Nitsch, M. O programa de biocombustíveis pro-álcool no contexto da estratégia energética brasileira. *Revista de Economia Política*. 1991, 11(2), 123-138.

Nunes, A.L.; Alves, T.W. Produção de biodiesel de dendê nos sistemas isolados do Amazonas: um meio para geração de desenvolvimento sócio-econômico para o estado. *I Simpósio Internacional de Oleaginosas Energéticas*. 2010. 173-177. Available from: http://www.cbmamona.com.br/pdfs/BID-02.pdf.

Ortega, M.C.; Moreno, M.T.; Ordovas, J.; Aguado, M.T. Behaviour of different horticultural species in phytotoxicity bioassays of bark substrates. *Scientia Horticulturae*. 1996, 66, 125- 132.

Pandey, A.; Soccol, C.R.; Nigam, P.; Soccol, V.T. Biotechnological potential of agro-residues. I: Sugarcane bagasse. *Bioresource Technology*. 2000, 74, 69-80.

Peng, F.; Peng, P.; Xu, F.; Sun, R.C. Fractional purification and bioconversion of hemicelluloses. *Biotechnology Advances*. 2012, 30, 879-903.

Peres, S.; Rangel, L.P.; Castelletti, E.; Almeida, C. *Estudo da viabilidade técnica para geração de energia elétrica a partir dos resíduos da mamona.* I Congresso Brasileiro de Mamona. 2004.

Piarpuzán, D.; Quintero, J.A.; Cardona, C.A. Empty fruit bunches from oil palm as a potential raw material for fuel ethanol production. *Biomass and Bioenergy.* 2011, 35, 1130-1137.

PNPB (2010) *Biodiesel.* Available from: http://www.biodiesel. gov.br /programa.html.

Roberto, I.C.; Felipe, M.G.A.; Lacis, L.S.; Silva, S.S.; Mancilha, I.M. Utilization of sugarcane bagasse hemicellulosic hydrolyzate by *Candida guiliermondii* for xylitol production. *Bioresource Technology.* 1991. 36(3), 271-275.

Rocha, C.J.A.; Silva, V.F.N.; Gonçalves, A.R.; Pimenta, M.T.B.; Rocha, G.J.M. Comparative study of chemical composition of lignocellulosic materials using analytical methodology developed for sugarcane bagasse. *33rd Symposiun on Biotechnology for Fuels and Chemicals.* 2011.

Rosa, S.E.S.; Garcia, J.L.F. O etanol de segunda geração: limites e oportunidades. *Revista do BNDES* 32. 2009, 117-156.

Sallet, C.L.; Alvim, A.M. Biocombustíveis: uma análise da evolução do biodiesel no Brasil. *Economia & Tecnologia.* 2011. 25, 1-13.

Santos, R.F.; Kouri, J. *Panorama Mundial do Agronegócio da Mamona.* 2° Congresso Brasileiro de Mamona. Embrapa Algodão. 2006.

Santos, S.F.; Bomfim, M.A.D.; Cândido, M.J.D.; Silva, M.M.C.; Pereira, L.P.S.; Neto, M.A.S. Effect of the castor bean hulls in the dairy goat diet on production, content and fat acid profile of milk. *Archivos Zootecnia.* 2011, 60, 113–122.

Satyanarayana, K.G.; Arizaga, G.G.C.; Wypych, F. Biodegradable composites based on lignocellulosic fibers – An overview. *Progress in Polymer Science.* 2009, 34(9), 982-1021.

Serrano, L.A.L.; Silva, C.M.M.; Ogliari, J.; Carvalho, A.J.C.; Marinho, C.S.; Detmann, E. Utilização de substrato composto por resíduos da agroindústria canavieira para produção de mudas de maracujazeiro-amarelo. *Revista Brasileira de Fruticultura.* 2006, 28(3), 487-491.

Severino, L.S. *O Que Sabemos Sobre o resíduo de mamona.* 2005. Available from: <www.cnpa.embrapa.br/publicacoes/2005/DOC134.pdf>.

Severino, L.S.; Auld, D.L.; Baldanzi, M. et al. A Review on the Challenges for Increased Production of Castor. *Agronomy Journal.* 2012. 104 (4), 853-880.

Severino, L.S., Moraes, C.R.A.; Gondim, T.M.S.; Cardoso, G.D.; Santos, J.W. Fatores de conversão do peso de cachos e frutos para peso de sementes de mamona. *Embrapa Algodão.* 2005, 15p.

Souza, A.D.V.; Fávaro, S.P.; Ítavo, L.C.V. et al., Caracterização química de sementes e tortas de pinhão manso, nabo-forrageiro e crambe. *Pesquisa Agropecuária Brasileira.* 2009. 44(10), 1328-1335.

Trzeciak, M.B.; Neves, M.B.; Vinholes, P.S.; Villela, F.A. Utilização de sementes de espécies de oleaginosas para produção de biodiesel. *Informativo ABRATES.* 2008, 18, 30-38.

Tsigie, Y.A.; Wang, C.Y.; Truong, C.T.; Ju, Y.H. Lipid production from *Yarrowia lipolytica* Po1g grown in sugarcane bagasse hydrolysate. *Bioresource Technology.* 2011, 102, 9216-9222.

Unica. 2012. Available from: http://english.unica.com.br/.

Vaz, P.H.P.M.; Sampaio, Y.S.B.; Sampaio, E.V.S.B. Análise da competitividade da mamona e da soja para produção de biodiesel no nordeste do Brasil. *Revista de Economia Agrícola*. 2010, 57(1), 35-48.

Wong, C.L.; Huang, C.C.; Chen, W.M.; Chang, J.S. Producing 2,3-butanediol from agricultural waste using an indigenuous *Klebsiella* sp. Zmd30 strain. *Biochemical Engineering Journal*. 2012, 69, 32-40.

Yates, N.L.; Rogers, M.N. Effects of time, temperature, and nitrogen source on the composting of hardwood bark for use as a plant growth medium. *Journal of the American Society of Horticultural Science*. 1981,106, 589-593.

Zhang, WL; Liu, ZY; Liu, Z; Li, FL. Butanol production from corncob residue using *Clostridium beijerinckii* NCIMB 8052. *Letters in Applied Microbiology*. 2012, 55(3), 240-246.

Zhu, Z; Zhu, M; Wu, W. Pretreatment of sugarcane bagasse with NH_4OH-H_2O_2 and ionic liquid for efficient hydrolysis and bioethanol production. *Bioresource Technology*. 2012, 119, 199-207.

In: Bioenergy Systems, Biological Sources … ISBN: 978-1-62417-330-1
Editor: Michel C. Allard

Chapter 5

HOW FARMING AND FORESTRY CONVERGE: ENHANCING THE INTERFACE BETWEEN AGRICULTURAL PRODUCTION, AND TREE BIOMASS/BIOENERGY SYSTEMS TO IMPROVE FARM-SCALE PRODUCTIVITY IN WESTERN AUSTRALIA

Mark P. McHenry*
School of Engineering and Energy,
Murdoch University, Western Australia

ABSTRACT

Shifting government policies and degrading terms of trade are leading to a range of tree biomass related options competing with food production on agricultural lands, including bioenergy, forestry, conservation legislation, and carbon biosequestration markets. Tree bioenergy plantations need not be monocultural and homogenous, and can be successfully incorporated into existing agro-ecological systems to increase primary productivity and food security while providing cost-effective bioenergy resources. This work examines tree integration in relation to wind speed, turbulence, humidity, evaporation, transpiration, temperature, water competition, solar use efficiency, frost, erosion, and fodder. These effects are examined in terms of farm scale limiting factors in agro-ecological systems (water, sunlight, wind, frosts, fodder, etc.), alongside research data on tree system integration with conventional livestock, horticultural and broadacre food production in southern Western Australia (WA).

* School of Engineering & Energy, Murdoch University, Western Australia. mpmchenry @gmail.com

Introduction

Agriculturalists require regionally-specific research demonstrating the physical costs and benefits of integrating new tree biomass industries with traditional primary production. New agroforestry opportunities and markets arising from climate change policies, natural resource management, and biodiversity conservation, promote a greater proportion of tree biomass cover in agricultural systems. While planting more trees on agricultural land seems to offer a simple solution to environmental and climate change objectives, agriculturalists must balance a range of options that compete for limited resources. It is unrealistic for large numbers of agriculturalists undertaking land-use change purely for environmental, carbon biosequestration, or biodiversity reasons (Fischer et al., 2008). Thus, this paper explores cases where trees have successfully (or otherwise) integrated into agricultural production systems to improve food production. By exploring cases where research shows simultaneous increases in farm productivity, and integrated tree-cover, progress can be made where macroeconomic environmental, social, and economic objectives overlap with the business of farming. Traditionally, trees supplied farmers with timber and fencing material at low cost (Gregory, 1995) and may have been a source of some net revenue from felled timber. As opposed to the commodity value of felled trees, their implicit value is revealed by their presence in agricultural landscapes worldwide (Manning et al., 2006). Trees clearly displace pasture and crops in the paddock. Yet there are economic reasons for this consistent pattern, especially as active replanting is labour intensive and land is at a premium.

Assessing integrated tree biomass return on investment is a complicated and cumbersome process, as the relative magnitudes of positive and negative effects of trees on a diverse range of primary crops, soils, rainfalls, seasons, and can generate a large range of current and future values. The recent upsurge in new biomass conversion technologies, carbon prices, and conservation policies add additional choices for diversification decision makers. This work recognises these uncertainties and focuses on the physical interactions between crops, animals and trees, to enable agriculturalists to adapt tree biomass systems for their particular climate, soil type, water resources, and business objectives. Due to the variability of agricultural land characteristics and the complexity and unpredictability of numerous natural cycles and their interactions, this work centres on research data derived from specific research plots over time (McNaughton et al., 1989; Sudmeyer and Speijers, 2007). The purpose of this work is to assist land allocation decision-making at the farm-scale, and critique the magnitude of both competition and synergies of tree system integration with cropping, livestock, and horticultural production systems.

Trees, Wind Speed, Microclimates, and Vegetative Competition

"Working trees" are most familiar to agriculturists as a windbreak to protect a primary crop (Perry et al., 2009). Single trees deflect wind around the tree, which reduces wind speeds near the tree, and correspondingly increases turbulence and wind speeds away from the tree (McNaughton et al., 1989; Lefroy et al., 2001). In simple terms, tree rows deflect winds upward to form a triangular zone of reduced wind speed from the crown of the windbreak line to the ground (Fig. 1) This deflection occurs at a horizontal distance roughly 2-5 times that of the windbreak height (H) windward, and 20-40 H in the lee, depending on the tree windbreak characteristics, wind speed, and the wind direction in relation to the tree-line. The effect on the wind changes when the wind direction changes relative to the tree-line (Sudmeyer and Scott, 2002). Sudmeyer and Scott (2002) found areas between 3 and 12 H leeward, the greatest wind speed reduction (almost 50%) occurred when the angle of incidence approached 90°, or perpendicular to the windbreak. However, when the angle of incidence of the wind was between 35 and 45° to the windbreak, the maximum wind speed reduction of almost 60%, occurred between 1 and 3 H in the lee of the tree row (Fig. 2). This demonstrates the variability of the magnitude of wind speed reduction in protected zones windward and leeward, with all possible angles of incidence (Sudmeyer and Scott, 2002) (Fig. 3). On a regional scale, the systematic planting of around 10% of land to shelterbelt trees could reduce local ground wind speeds by 50% (Bird et al., 1992).

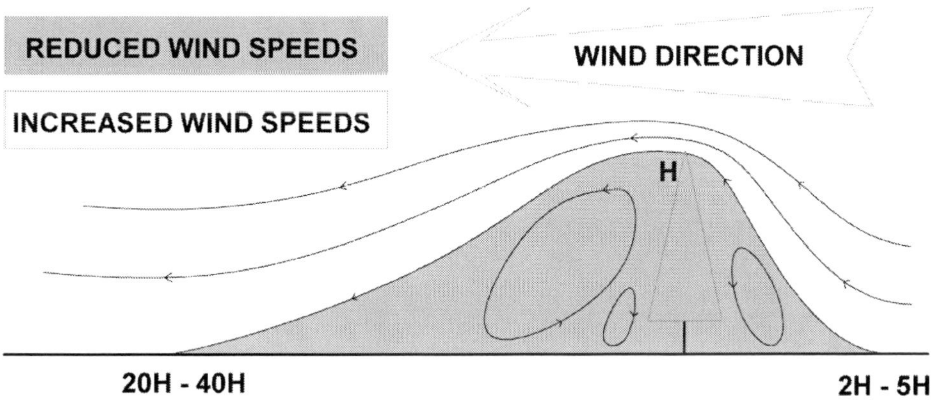

Figure 1. Two-dimensional tree schematic showing areas of increased wind speeds and windward and leeward in terms of windbreak height (H). The horizontal and vertical scales are not to scale.

The porosity of a windbreak is an important determinant of the effect it will have on crops. Porosity is an optical measure defined as the ratio of open to total area, as seen from the wind direction (Judd et al., 1996). Windbreaks with porosities of between 70-30% (a porous windbreak), will lead to wind speed reductions that are inversely proportional to the windbreak density. For example, a windbreak with a porosity of 60% reduces wind speeds by around 40%, while a relatively dense windbreak of 30% porosity reduces wind speeds by around 70%. The higher the windbreak porosity, the smaller the effect on wind speed, air temperature, and humidity (Cleugh et al., 2002). However, at extremes of porosity, dense windbreak of porosities less than 20 can induce eddies in the immediate lee (0-10 H), which increases turbulence encountered by crops. The leeward wind speed reduction for very low porosity windbreaks is less than medium porosity windbreaks (Vigiak et al., 2003). Therefore, the most effective windbreak porosity is around 35%, as it provides the best shelter over long, intermediate and short distances, without creating excessive circulation and turbulence leeward of the windbreak (Santiago et al., 2007).

Designing a windbreak system with the optimum porosity is not a simple concept in practice, as the shape and density of individual trees vary from top to bottom, and between each tree in the line (McNaughton et al., 1989). The windbreak porosity also varies throughout the year due to seasonal leaf changes, especially with deciduous trees. Seasonal porosity changes may be either beneficial or detrimental to the primary crop, and the choice of windbreak design, and tree species, is crucial to achieve the required wind speeds for their production system.

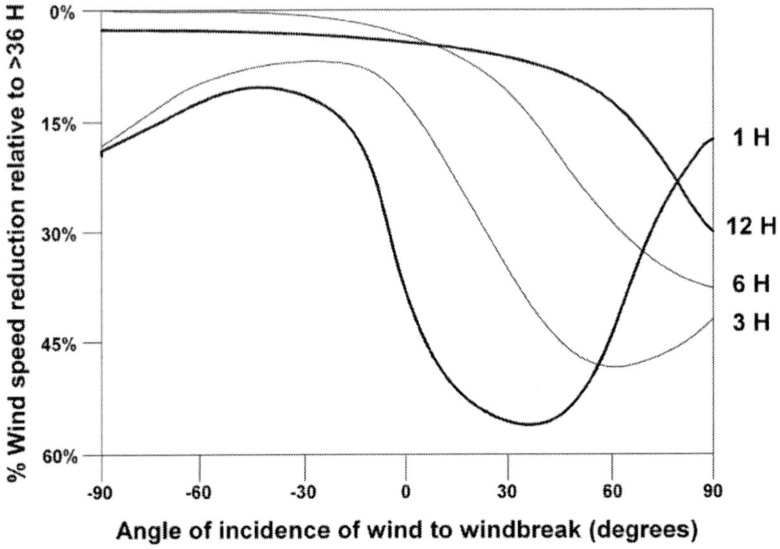

Figure 2. A stylised chart of 1996 results from Sudmeyer and Scott (2002) of wind speed and angles of incidence at positions windward and leeward of a windbreak at their Howick research site. Positive angles indicate leeward measurements. Negative angles represent windward data.

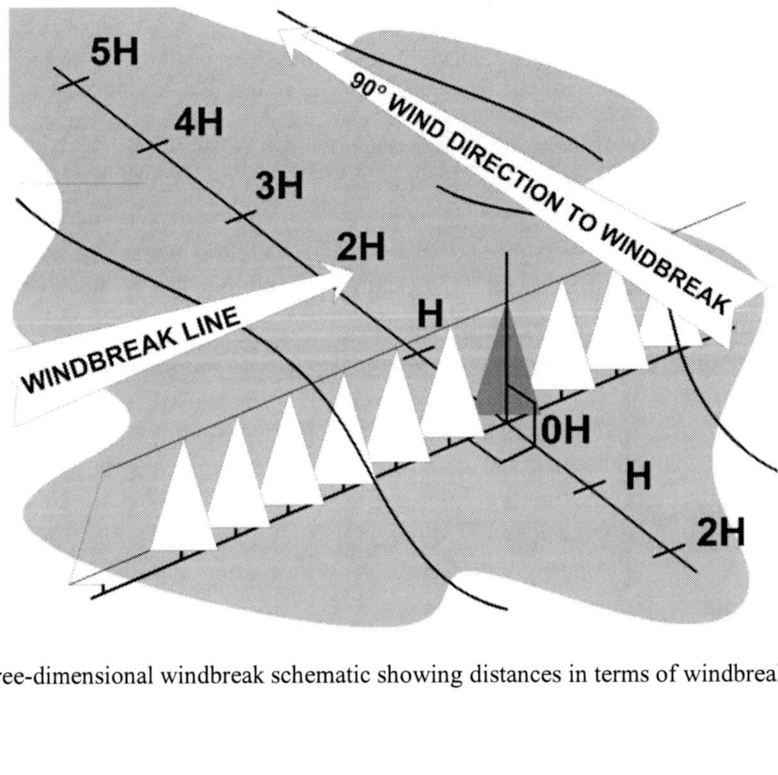

Figure 3. Three-dimensional windbreak schematic showing distances in terms of windbreak height (H).

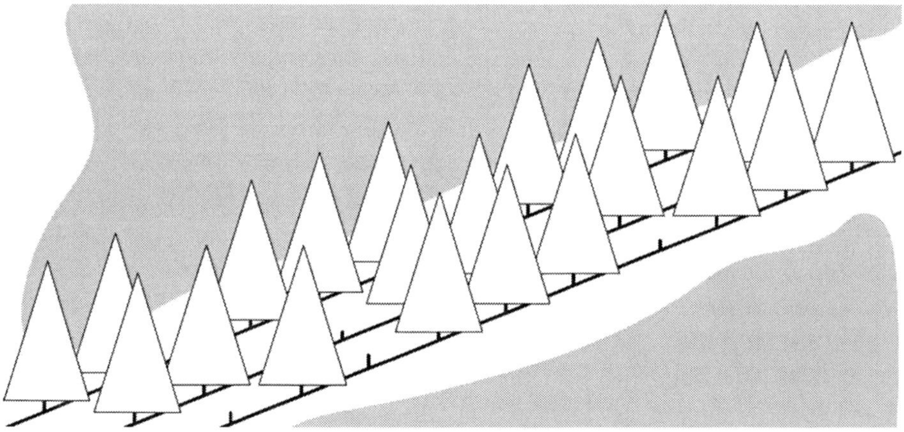

Figure 4. Three-dimensional windbreak schematic showing no significant gaps, even with a survival rate of 70%.

Windbreaks should ideally be consistent along the line to maximise wind protection, as the majority of the reduction in wind speed depends on tree height and porosity (McNaughton et al., 1989; Cleugh and Hughes, 2002). Gaps in windbreaks result in acceleration of wind speeds through these areas, and can cause erosion and damage to crops (Cleugh et al., 2002; Sudmeyer and Scott, 2002). Using multiple rows of trees together in a narrow design will reduce the likelihood of gaps appearing due to tree injury, death, or disease (Hennon and McClellan, 2003) (Fig. 4).

Reducing wind speeds alters heat and water vapour fluxes in the air (Cleugh, 2002). As turbulence carries heat and vapour, it can determine which parts of a paddock are warm, cool, humid, or dry (McNaughton et al., 1989). The use of windbreaks can also reduce sensible heat advection (the transfer of heat energy by means of horizontal mass motion), which reduces water evaporation, especially during irrigation, and in dry years (Bird, 1998). In the vast majority of cases, this is an advantage as the conservation of soil moisture improves plant growth. However, this can also contribute to a higher incidence of disease in a minority of crops when the evaporation from leaf surfaces is reduced alongside increased humidity. This is more likely to occur with irrigated crops (Brandle et al., 2004). Sudmeyer and Scott (2002) recorded a summer season of wind speed, relative humidity, and air temperature when the wind angle of incidence was 90° to the windbreak. They found that the leeward wind speed decreased between 11-48% within 20 H, with the greatest reduction in wind speed occurring between 3 and 6 H. The air temperature and relative humidity in areas within 6 H were both increased up to 7%, relative to unsheltered areas. Between 6 H and 12 H the air temperature and relative humidity increase was only 2%, compared to unsheltered areas beyond 12 H. While these changes in relative humidity and air temperature are small, the majority of the significant effects on the reduction in wind speed are largely confined to the quiet zone, which is generally between 2 and 8 H (Cleugh et al., 2002; Sudmeyer and Scott, 2002). While much windbreak literature has slight differences in definitions of the quiet zone, alongside numerous micro-meteorological empirical and modelled data, much uncertainty remains in the net benefit to agricultural crop productivity (McNaughton et al., 1989).

An extensive Australian windbreak study (concluding in 2002) known as the Australian National Windbreak (ANW) program, found only relatively small changes in the paddock microclimate due to windbreak integration. This was suggested to be due to experimental plots with single line windbreaks, which left crops exposed to day to day and seasonal variability of wind direction. The extensive analyses undertaken under the ANW program recommended either closely spaced windbreaks, or those that surround the perimeter of paddocks to create consistent protection over a greater number of wind directions (Cleugh et al., 2002). This advice came with a caveat that increased tree plantings may reduce yields in the competition zone, a very important factor in Australian wind break design (Oliver et al., 2005) (Fig. 5).

The zone of competition reduces crop yields due to the trees and crop competing for limited moisture, nutrients, light, and also allelopathic effects of some trees (Sudmeyer et al., 2002; Oliver et al., 2005). The type and magnitude of competition depends on the crop, windbreak species, the soil type, and the climate. Sudmeyer and Scott (2000) found that within 3 H of Maritime pine (*Pinus pinaster*) windbreaks in the WA Howick research site, (100 km east of Esperance), the amount of stored soil water was reduced by 100-150 mm, up to 1.8 m of depth.

This halved the water available for crop transpiration, compared to distances greater than 3 H from the pines. The negative effects on crops were more pronounced in drier years within 3 H. Sudmeyer et al. (2002) found that pruning tree roots reduced the expanse and magnitude of crop losses next to trees where roots are confined close to the surface. Competition between the primary crop and the windbreak can be minimised by selecting taller trees to be planted in narrow windbreaks (around 3 tree rows), positioned at least 30 H apart. The tree roots should be deep-ripped periodically to reduce the competition zone (Sudmeyer et al., 2002).

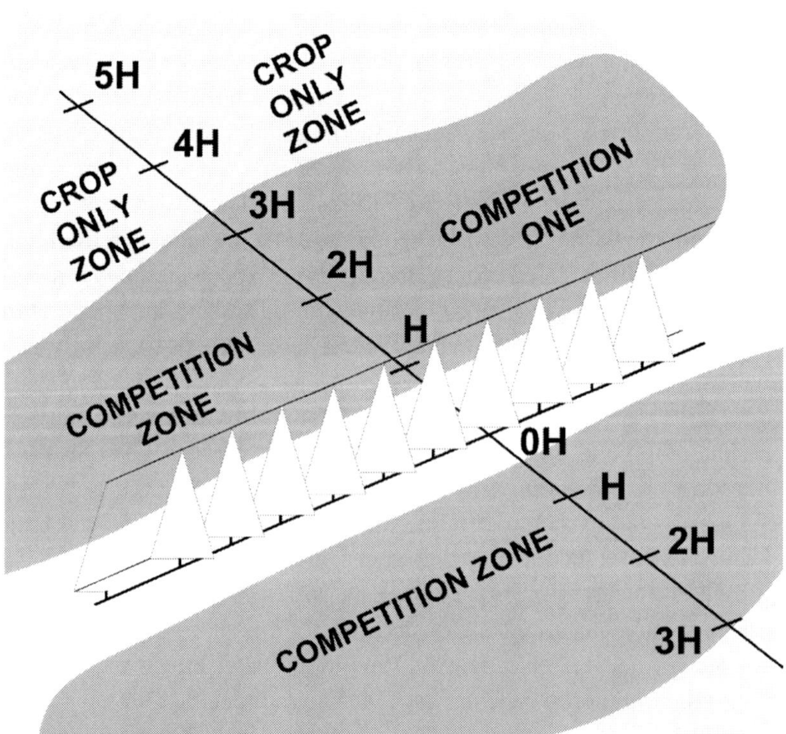

Figure 5. Windbreak schematic and the competition zones between trees and crops.

WINDBREAKS, CROPS, ENERGY, AND WATER USE EFFICIENCY

As evaporation is proportional to the square root of the wind speed, a reduction in wind speed will reduce evaporative water loss (Sudmeyer and Scott, 2002). Windbreaks that reduce wind speeds in the early growing season increase chances of seed germination, while at the end of the growing season the reduced water loss can increase water use efficiency, and prolong the season (Sudmeyer et al., 2002). Research undertaken by Campi et al. (2009) in a European Mediterranean region found wheat crop evapotranspiration reduced by up to 16% behind a Arizona cypress (*Cupressus arizonica* L.) windbreak, with a corresponding temperature increase of 9% in the areas of maximum wind protection (1.3 - 4.7 H). ANW studies by Sudmeyer and Scott in 2002 showed that at the Howick site, the average potential soil evaporation was 20% less than areas with no windbreak. The ANW program summary estimated that crop evapotranspiration was reduced by less than 3% in the lee of windbreaks, alongside decreased evaporation, suggesting increased crop transpiration (Sudmeyer and Scott, 2002). In areas of high turbulence, evaporation rates are greater than in areas within the quiet zone. When there are reduced levels of evaporation in wind protected areas, the stomata of some plants may remain open and maintain high transpiration rates from increased photosynthesis, potentially increasing evapotranspiration rates (Bird, 1998). The shelter provided by windbreaks can sometimes have little effect in practice on transpiration from crops, even when the levels of evaporation are reduced. This is because the amount of water transpired by plants depends on the availability of water in the soil and other factors, such as air humidity windward of the area (Cleugh, 2002; Sudmeyer and Scott, 2002). The interpretations of evaporation, transpiration and evapotranspiration processes in the lee of windbreaks are complex, and often simultaneously increase and decrease water loss and retention in the paddock. Crop surface tissue damage from wind increases transpiration, and windbreak protection can lead to circumstances of reduced crop transpiration. However, protection can also lead to increased transpiration levels due to plants growing larger leaves (Bird, 1998). Therefore, the measured levels of crop transpiration and evapotranspiration can be misinterpreted, unless the origins for such increases or decreases are known alongside yield.

An additional means of increasing water use efficiency is by using perennial plants to increase the physical infiltration of rainwater. Runoff is substantially reduced under trees and shrubs, due to decayed root channels, holes made by soil fauna, higher organic matter content, and general soil porosity (Cleugh et al., 2002). Nulsen et al. (1986) reported in a southwest Australian study that 25% of rain falling on the canopy of mallee[1] trees ran down the stem and was directed around the bole to soil root channels at depths of 28m, to apparently be used in the dry season. The reduction of groundwater salinity and water-logging to adjacent paddocks is a critical issue in Australian agriculture. However, there are very specific conditions required for trees to reduce groundwater levels. Trees are more effective at lowering the water table when the water is fresh, and aquifers are small/local, rather than regional. Groundwater reduction by trees rarely exceeds 30 m from their base (George et al., 1999; Hall et al., 2002). Therefore, planting rows at 60m intervals would likely reduce the water table, albeit at the expense of yields with such proximal distances the tree-crop

[1] a general term used to describe multi-stemmed, small species of Eucalypt

interface (Oliver et al., 2005). A study of 80 tree planting sites in the southwest of WA by George et al. (1999) on reducing groundwater levels and salinity encroachment, found that the trees will only make a significant difference in a large regional aquifer if there are considerable areas of the catchment (70-80%) planted to trees.

In addition to water use efficiency, the design of windbreak systems can increase the energy use efficiency of a farm. This can be achieved by dispersing trees with differing canopy characteristics with crops or pasture arranged and orientated to intercept solar energy at different canopy levels, in alternate seasons (Patabendige et al., 1992). This can increase the net primary production considerably from the difference in active growth periods between annuals and perennials. In the higher rainfall areas of southwest WA, annuals active growth periods are usually between five and seven months, whereas perennials can extend the growing season and take advantage of summer and opening rains (Patabendige et al., 1992; McDowall et al., 2003).

Windbreak Trees, Horticultural Farm Yields, and Frosts

Fruits and vegetables are often more sensitive to wind stress and show yield and quality reductions in even moderate winds. Shelter and climate moderation for horticultural production contributes to increased marketable yield and individual fruit weight (Brandle et al., 2004). Protection at critical times, such as when seedlings are young, or during flowering periods, can increase plant survival and improve fruit set by reducing damage from abrasive winds (Cleugh et al., 2002; Brandle et al., 2004). Using windbreaks in high-value horticultural areas can reduce wind speeds for wind pollinated crops, provide floral species for maintenance of insect and bird pollinator colonies throughout the year, as well as attract parasitic wasps, or other predatory insects and birds. However, providing some types of trees in shelterbelts can cause problems. Plants that produce heavy nectar can attract some birds that attack flowering crops, and by providing niches where smaller birds can peck fruits without corresponding niches for their predators (Yunusa et al., 2002).

Some crops grown in sheltered areas also tend to mature earlier than unsheltered crops. This is often due to the several degree increase in daytime temperatures for temperate regions from the reducing turbulent mixing, leading to a slight increase (1 or 2°C) at night near ground level. In temperate regions these increases in temperature usually appear earlier in the growing season (Brandle et al., 2004). This can be an advantage for crops such as fruit and vegetables that receive premium prices at particular times of the year, in addition to potentially expanding the range of alternative crop cultivars. The regulation of soil temperatures by windbreaks may reduce the losses from winter and spring frosts in southwest WA. The common type of frost in the southwest is called a radiation frost. On clear and calm nights, radiative heat losses by soil and vegetation can reduce the air temperature to below 0°C, resulting in the condensation on plant surfaces freezing to produce a frost (Brandle et al., 2004). A slightly increased soil temperature will protect against some radiation frost, but this effect is likely to be small.

TREES, LIVESTOCK, FODDER TREES AND PERENNIAL PASTURES

There are many documented benefits windbreaks can offer animal husbandry including: improved growth rates; increased ovulation rates in both sheep and cattle; increased wool growth rate; reduced lamb mortality, and; decreased abortions rates induced by hypothermia (Gregory, 1995). Research undertaken by Holmes et al. (1978) in New Zealand, found that 150 kg dairy heifers grew faster than control twins when they were provided access to shelter in their paddocks (Gregory, 1995). Windbreaks also offer protection from the sun in hot conditions, which has been shown to improve milk yield, milkfat yield, assist the prevention of mastitis, improve conception rates in dairy cattle, and increase growth rates in fattening cattle (Gregory, 1995; Mader, 2003). Altering the microclimate for stock protection is vital in many regions of the southwest of WA, especially during winter storms and on hot summer days (Pollard et al., 1999; Cleugh et al., 2002; Mader, 2003). Winter born lambs and shorn sheep are especially at risk from cold, wet and windy conditions (Gregory, 1995; Pollard et al., 1999). Exposure to high wind speeds increases the lower critical temperature for animals, which is wasteful in terms of energy utilisation (Gregory, 1995).

Extremes of weather cause much animal discomfort in confined areas, and the provision of shelter allows livestock to self-regulate to improve their individual welfare. The design of windbreaks for livestock can also incorporate the preferences of the animals under critical periods (Pollard et al., 1999). For example, many ewes prefer to lamb in isolation from the rest of the flock, and also have a tendency to lamb at the perimeter of the paddock. In cold winter conditions when flocks huddle together behind windbreaks, lambing ewes may stray from the flock and increase their environmental exposure. Therefore, providing scattered blocks of trees in paddocks is more than thin windbreaks for lambing paddocks (Prinsley, 1992; Gregory, 1995).

In most regions, extreme weather events are unlikely to occur every year, but windbreaks can be justified as a form of stock insurance (Sudmeyer et al., 2002). It was suggested in a 1973 study by Sturrock in New Zealand that for optimum protection of livestock, a minimum of 5% of land under tree cover was required (Gregory, 1995). However, the farmer would realise that from a productive efficiency perspective, the reduced pasture available to animals with the introduction of inedible windbreaks may cancel out the benefits of shelter for animals (da Silva et al., 1993). Edible windbreaks can compensate for such pasture loss, especially as a summer source of fodder (Cleugh et al., 2002). The lack of additional green feed in late summer and autumn is a major constraint to livestock production in the southwest of WA, and importing feed is particularly expensive in years with region-wide feed shortages. The ability of some tree species to provide fodder and forage to supplement seasonal food shortages, defer the grazing of annual pastures, and reduce hand feeding, has generated much interest in finding suitable species for specific production systems and climates (Patabendige et al., 1992; Cleugh et al., 2002).

Two tree species that have received much attention for use as forage in WA are the tagasaste (*Chamaecytisis palmensis*) and saltbushes (*Atriplex spp.*). Tagasaste is suited to well-drained soils in areas of higher than 400mm of annual rainfall, and saltbushes are tolerant to moderately saline areas and occasional waterlogging, but is often used in dryer climates. Both are sometimes used as sheep feed in the autumn feed-gap and provide vegetative cover on erosion-prone sites (Patabendige et al., 1992; de Koning and Milthorpe, 2008). Research by de Koning and Milthorpe (2008) reviewed the use of various saltbush species as a fodder crop, and found that when grown in salty or moderately saline affected areas, the feed value of these species are poor without ample fresh drinking water. The study recommended the use of saltbush species for fodder only from areas with fresh groundwater, or solely for environmental reclamation of salt affected land (de Koning and Milthorpe, 2008). While tagasaste has proved itself as a high value fodder crop in areas of WA with lower rainfall, some tree designs create a level of pasture/crop competition that decreases farm productivity (Lefroy et al., 2001; de Koning and Milthorpe, 2008). Integrating fodder plantations into parallel rows in paddocks (also known as alley-cropping), have been a relatively popular design, although it requires quantification of various levels of competition between the tree alley, and the pasture/crop (Lefroy et al., 2001). An alley-cropping and water balance study on deep sand in Moora, by Lefroy et al. (2001), compared control yields with three integrated crops, using another species of tagasaste (*C. proliferus*). Trees were spaced at 30m between rows, and 0.7m between each tree. Experiments with the two lupin varieties (*Lupinus angustifolius* vars. Gungurru and Merrit)), and the oats (*Avena sativa* var. Toodyay) found that segregated monocultures performed better than alley cropping. In the first year, the Gungurru lupins provided a yield increase of 23%, including the land area under the four year old tagasaste trees pruned from 2m to 0.6m in year zero of the trial. However, the second year Toodyay oat and third-year Merrit lupin crops suffered significant losses of 8 and 32%, respectively. The oat crop losses were due solely to the trees displacing the crop aboveground, with no below-ground competition verified by pot trials. However, the continuing tagasaste tree growth both displaced and competed with the second variety of lupins, at a magnitude of 20 and 12%, respectively. Lefroy et al. (2001) offered a review and recommendation to agroforestry literature of segregating trees from crops, and targeting landscape niches when production is water-limited.

If alley cropping is a preferred option, such productivity losses may be offset in economic terms by the tagasaste fodder eaten by stock in the "feed gap" period (Lefroy et al., 2001). A study undertaken by Scott (1990) used tagasaste as a management strategy to improve the efficiency of sheep production on a Badgingarra property (30km northwest of Moora), over the summer and autumn feed-gap period, and compared it to grain and dry-fed animals. The research found that in addition to greater wool production over the summer period, the 8% discounted economic analysis suggested that feeding the sheep tagasaste over the summer feed-gap period would become cost effective in the sixth year after planting. This analysis included the lost production from a segregated paddock-style plantation, and the additional labour costs of establishing the plots. Other trees that have received attention for forage value in the southwest are the golden wreath wattle (*Acacia saligna*), the mulga (*Acacia aneura*) the black sheoak (*Casurina cristate*), the swamp sheoak (*Casurina obesa*), carob (*Ceretonia siliqua*), honey locust (*Gleditsia trichanthos*), and leucaena (*Leucaena leucocephala*), in addition to various other wattles, poplars, tamarisks, and willows. However, some wattles have uncertain palatability and digestibility for sheep and cattle, due to high leaf-tannin content (Patabendige et al., 1992; Lefroy et al., 2001). A significant issue with the introduction of some fodder tree species is their potential to become an invasive pest (Lefroy et al., 2001). As an alternative to perennial tree fodder species, perennial pastures offer another option for farmers to supplement stock diets in the feed-gap period. A study by Sanford et al. (2003) focused on the importance of using perennial pastures to fill the feed-gap with non-fodder windbreaks on both Merino sheep and salinity. The study measured sheep carrying capacity, wool production and groundwater recharge in the southwest of WA near Albany. The study found that in the growing season, there was similar herbage in the annual pasture consisting mostly of subterranean clover (*Trifolium subterraneum*), and erodium (*Erodium botrys*), and the perennial kikuyu pasture (*Pennisetum clandinestinum*), with small amounts of subterranean clover. The annual pasture stocking rate was 12 dry sheep equivalents per hectare (DSE ha^{-1}), and the kikuyu was 14 DSE ha^{-1}. In spring, all stock rates were increased, and in autumn all paddocks were destocked. In summer and autumn, the kikuyu paddock contained between 350 and 4,900kg more dry matter per hectare (DM ha^{-1}) than the annual pasture, mostly due to summer rains. Both the carrying capacity and clean wool production per hectare was significantly higher on kikuyu pasture in the years with summer rainfall (Sanford et al., 2003). These control paddocks were compared with paddocks with ten-year-old blue gum tree belts (*Eucalyptus globulus*), planted at 1250 ha^{-1}, with 4m between rows and 2m tree spacing. The adjacent annual pasture suffered reduced herbage accumulation in the paddock with tree belts by 16%, while not affecting the perennial kikuyu herbage accumulation. The research found that the tree competition of the closely spaced native evergreen blue gum windbreaks reduced the carrying capacity of the annual and kikuyu pasture by an average of 10%, and the annual average clean wool production by 13%. The research concluded that there was less tree-pasture competition with kikuyu pastures than annual pastures. The research showed that kikuyu pasture, without trees produced significant increases in livestock production, in addition to substantial groundwater reductions, relative to the annual pastures. Sanford et al. (2003) recommended a combination of balancing kikuyu and trees to lower the water table, and to slightly lower livestock production.

Trees and Broadacre Yields

There is considerable variation in response to shelter from common broadacre crops (Sudmeyer et al., 2002; Sudmeyer and Speijers, 2007). Much of this variation is due to the growth habits, and different positions of the shoot apex. If the shoot apex is carried at the top of the plant canopy, the plant is more likely to be sensitive to wind damage (Sudmeyer et al., 2002). In Australian agricultural systems, crop yields exhibit large annual variations. Significant yield differences occur between crop varieties, regions, and even across paddocks (Sudmeyer and Scott, 2002). Notwithstanding such difficulties, crop research in Esperance undertaken by Sudmeyer et al (2002), integrated various pine, Eucalypt timber belts, and remnant mallee vegetation with broadacre crops. The results suggested that the sheltered primary crop yield at 5-20 H was slightly greater than the unsheltered yield at 20-30 H, in similar soil and topographical conditions. These increases were only statistically significant at 9 H, and must be evaluated against significantly reduced yields within 4 H of the trees. These trials included crops of barley (*Hordeum vulgare*), wheat (*Triticum aestivum*), lupins (*Lupinus angustifolius*), canola (*Brassica napus*), and one trial of faba bean (*Vicia faba*) (Sudmeyer et al., 2002). Sudmeyer et al. (2002) stated that the yields of the 37 windbreak sites on the south coast of WA in years with average or above-average rainfall between 4 and 20 H were generally similar to unsheltered broadacre yields. This contrasted with mean yields between 1 and 20 H reduced by 2-6%. In dry years, when crops received less than 50% of the average annual rainfall, the mean sheltered yield between 5 and 20 H was 11% greater than the unsheltered yield, although the mean yield between 1 and 20 H being similar to the unsheltered yield (Sudmeyer et al., 2002). The greatest improvements in crop yield were recorded in 1996 at Jerdacuttup (around 120km west of Esperance), where an estimated 20% of the cropping areas were severely effected by wind erosion. In this year the tree windbreaks increased the mean yield of wheat and barley between 1 and 20 H by 25% and reduced sandblasting damage as far as 40 H, beyond which some crops were completely destroyed (Sudmeyer et al., 2002). The research showed that within 4 H, yields decreased with increasing proximity to the trees, although this varied with the amount of rainfall received in the growing season. In years with average or above-average rainfall, the yield decreased out to 2-4 H from the trees, and the mean yield between 1 and 4 H was 17-26% less than the unsheltered yield. In the dry years, the yield was reduced out to 3-5 H from the trees, and the mean yield between 1 and 5 H was 35% less than unsheltered yield (Sudmeyer et al., 2002). The overall results from the study were that the windbreaks decreased sheltered (1-20 H) crop yields in areas that suffer severe wind erosion by 1% compared to unsheltered areas (20-30 H) in the 64 field-years of the study between 1994 and 1997. This compared against a decrease of 4% in areas that do not suffer severe wind erosion in 58 field-years over the same interval. Sudmeyer et al. (2002) concluded that the net yield between 1 and 20 H was similar to the yield of the unsheltered crops over time, due to sheltered zones improving relative yields in dry and windy years, which offset reduced yields in the competition zone.

An economic analysis undertaken by Jones and Sudmeyer (2002) on the data from the 37 sites found that windbreaks reduce net agricultural income when the primary crop does not suffer wind damage. However, productivity increases in the lee of windbreaks in this region of WA could offset yield losses and the cost of establishing trees if unsheltered crops were subject to 3 or 4 severe wind damage events over a 35 year interval. The analysis found that appropriately spaced windbreaks in the region were unlikely to reduce the net present value of crop production more than 6.5% over the long term. Any net financial benefits from windbreaks are dependent on the timing, and frequency of wind damage in relation to the age, size and distance between the windbreaks, the extent of damage that normally occurs due to wind, and whether or not the lateral roots of trees are pruned (Jones and Sudmeyer, 2002). The primacy of below-ground windbreak-crop root competition, and the benefits of wind protection, contrasts with the practice of orientating tree lines north-south to reduce tree shading impacts. Some crops such as wheat will benefit early in the season from north-south orientation, although other crops such as lupins will not. Therefore, depending on the crop, and whether wind, water or light is the yield limiting factor, windbreak orientations should generally be orientated to protect from the common direction of damaging winds at sensitive times of the year (Sudmeyer and Speijers, 2007; Campi et al., 2009) (See Fig. 6 and 7 for an example of seasonal Albany wind frequency analyses (wind roses) for Summer, Autumn, Spring, and Winter at 9am, and 3pm, respectively).

The effects of windbreaks on broadacre crops have also been examined under climate scenarios to model the potential of windbreaks to protect dryland maize from the negative effects of climate change. A study in the USA by Easterling et al. (1996) found that shelter increased dryland yields relative to unsheltered crop yields for almost all levels of climate change. This was due to the windbreak providing night-time cooling, relieving some water stress which compensated for a shorter growing season. The most positive effect of shelter was for the most severe cases of large losses in rainfall, and large wind speed increases

(Easterling et al., 1996). These and other similar studies suggest a greater role for windbreaks in moderating future climatic changes in the vulnerable southwest of WA.

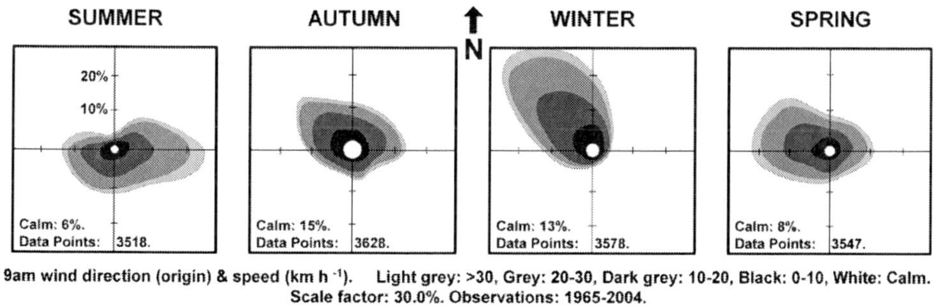

Figure 6. Seasonal 9am wind roses for Albany airport. (Data derived from the Bureau of Meteorology).

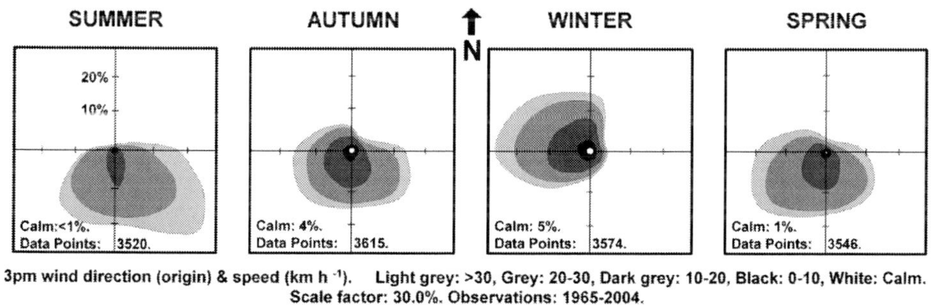

Figure 7. Seasonal 3pm wind roses for Albany airport (Data derived from the Bureau of Meteorology).

CONCLUSION

Appropriate agricultural windbreak configurations allow farmers to increase the productivity, and profitability of their farming enterprise. With carefully planned tree biomass systems, agriculturalists have the potential to increase yields, improve product quality, and reduce losses during extreme-wind conditions. Harnessing tree systems that protect crops, pastures and livestock in key seasons can be an inexpensive form of insurance for agriculturalists. However, when planting trees for either conventional, strategic, forestry, or ecological objectives, farm productivity may be inadvertently reduced with regionally and seasonally inappropriate tree systems.

ACKNOWLEDGMENT

Many thanks, once again, to Julia Anwar McHenry.

REFERENCES

Bird, P.R. 1998. Tree windbreaks and shelter benefits to pasture in temperate grazing systems. *Agroforestry Systems*. 41: 35-54.

Bird, P.R., D. Bicknell, P.A. Bulman, J.A. Burke, J.F. Leys, J.N. Parker and F.J. Van Der Sommen. 1992. The role of shelter in Australia for protecting soils, plants and livestock. *Agroforestry Systems*. 20: 59-86.

Brandle, J.R., L. Hodges and X.H. Zhou. 2004. Windbreaks in North American agricultural systems. *Agroforestry Systems*. 61: 65-78.

Campi, P., A.D. Palumbo and M. Mastrorilli. 2009. Effects of tree windbreak on microclimate and wheat productivity in a Mediterranean environment. *European Journal of Agronomy*. 30: 220-227.

Cleugh, H.A. 2002. Parameterising the impact of shelter on crop microclimates and evaporation fluxes. *Australian Journal of Experimental Agriculture*. 42: 859-874.

Cleugh, H.A. and D.E. Hughes. 2002. Impact of shelter on crop microclimates: a synthesis of results from wind tunnel and field experiments. *Australian Journal of Experimental Agriculture*. 42: 679-701.

Cleugh, H.A., R. Prinsley, R.P. Bird, S.J. Brooks, P.S. Carberry, M.C. Crawford, T.T. Jackson, H. Meinke, S.J. Mylius, I.K. Nuberg, R.A. Sudmeyer and A.J. Wright. 2002. The Australian National Windbreaks Program: overview and summary of results. *Australian Journal of Experimental Agriculture*. 42: 649-664.

Easterling, W.E., C.J. Hays, M. McKenney and J.R. Brandle. 1996. Modelling the effect of shelterbelts on maize productivity under climate change: An application of the EPIC model. *Agriculture, Ecosystems and Environment*. 61: 163-176.

Fischer, J., B. Brosi, G.C. Daily, P.R. Ehrlich, R. Goldman, J. Goldstein, D.B. Lindenmayer, A.D. Manning, H.A. Mooney, L. Pejchar and J. Ranganathan. 2008. Should agricultural policies encourage land sparing or wildlife-friendly farming? *Frontiers in Ecology and the Environment*. 6: 380-385.

George, R. J., R.A. Nulsen, R. Ferdowsian and G.P. Raper. 1999. Interactions between trees and groundwaters in recharge areas - A survey of Western Australian sites. *Agricultural Water Management*. 39: 91-113.

Gregory, N.G. 1995. The role of shelterbelts in protecting livestock: a review. *New Zealand Journal of Agricultural Research*. 38: 423-450.

Hall, D.J.M., R.A. Sudmeyer, C.K. McLernon and R.J. Short. 2002. Characterisation of a windbreak system on the south coast of Western Australia. 3. Soil water and hydrology. *Australian Journal of Experimental Agriculture*. 42: 729-738.

Hennon, P.E. and M.H. McClellan. 2003. Tree mortality and forest structure in the temperate rainforests of southeast Alaska. *Canadian Journal of Forest Research*. 33: 1621-1634.

Judd, M.J., M.R. Raupach and J.J. Finnigan. 1996. A wind tunnel study of turbulent flow around single and multiple windbreaks, Part I: Velocity fields. *Boundary-Layer Meteorology*. 80: 127-165.

Jones, H.R. and R.A. Sudmeyer. 2002. Economic assessment of windbreaks on the south-eastern coast of Western Australia. *Australian Journal of Experimental Agriculture*. 42: 751-761.

de Koning, C., and P. Milthorpe. 2008. Integrating forage shrubs: their potential in mixed-farming enterprises in low rainfall regions Union Offset. Canberra, Australia. (*http://rirdc.infoservices.com.au/downloads/08-044.pdf*).

Lefroy, E.C., R.J. Stirzaker and J.S. Pate. 2001. The influence of tagasaste (*Chamaecytisus proliferus* Link.) trees on the water balance of an alley cropping system on deep sand in south-western Australia. *Australian Journal of Agricultural Research*. 52: 235-246.

Mader, T.L. 2003. Environmental stress in confined beef cattle. *American Society of Animal Science*. 81: E110-E119.

Manning, A.D., J. Fischer, and D.B Lindenmayer. 2006. Scattered trees are keystone structures - Implications for conservation. *Biological Conservation*. 132: 311-321.

McDowall, M. M., D.J.M. Hall, D.A. Johnson, J. Bowyer and P. Spicer. 2003. Kikuyu and annual pasture: a characterisation of a productive and sustainable beef production system on the South Coast of Western Australia. *Australian Journal of Experimental Agriculture*. 43: 769-783.

McNaughton, K.G., M.H. Unsworth and M.R. Raupach. 1989. Micrometeorology of Shelter Belts and Forest Edges [and Discussion]. *Philosophical Transactions of the Royal Society of London*. 324: 351-368.

Nulsen, R.A., K.J. Bligh, I.N. Baxter, E.J. Solin and D.H Imrie. 1986. The fate of rainfall in a mallee and heath vegetated catchment in southern Western Australia. *Australian Journal of Ecology*. 11: 361-371.

Oliver, Y.M., E.C. Lefroy, R. Stirzaker and C.L. Davies. 2005. Deep-drainage control and yield: the trade-off between trees and crops in agroforestry systems in the medium to low rainfall areas of Australia. *Australian Journal of Agricultural Research*. 56: 1011-1026.

Patabendige, D.M,. P.R. Scott and E.C. Lefroy. 1992. Fodder trees and shrubs for high rainfall areas of south Western Australia. *Resource Management Technical Report No. 135*. South Perth, Western Australia: Department of Agriculture, Western Australia.

Perry, C.H., C.W. Woodall, G.C. Liknes and M.M. Schoeneberger. 2009. Filling the gap: improving estimates of working tree resources in agricultural landscapes. *Agroforestry Systems*. 75: 91-101.

Pollard, J.C., K.J. Shaw and R.P. Littlejohn. 1999. A note on sheltering behavior by ewes before and after lambing. *Applied Animal Behaviour Science*. 61: 313-318.

Prinsley, R.T. 1992. The role of trees in sustainable agriculture - an overview. *Agroforestry Systems*. 20: 87-115.

Sanford, P., X. Wang, K.D. Greathead, J.H. Gladman and J. Speijers. 2003. Impact of Tasmanian blue gum belts and kikuyu-based pasture on sheep production and groundwater recharge in south-western Western Australia. *Australian Journal of Experimental Agriculture*. 43: 755-767.

Santiago, J.L., F. Martin, A. Cuerva and N. Bezdenejnykh. 2007. Experimental and numerical study of wind flow behind windbreaks. *Atmospheric Environment*. 41: 6406-6420.

Scott, P.R. 1990. Agroforestry - Integration of trees into the agricultural landscape. *Resource Management Technical Report No. 102*, South Perth, Western Australia: Western Australian Department of Agriculture.

da Silva, R.G., S. Huda and J. Flesch. 1993. Weather and climate and animal production. In: *Guide to Agricultural Meteorological Practices*. WMO No. 134. Geneva: World Meteorological Organization.

Sudmeyer, R.A. and P.R. Scott. 2002. Characterisation of a windbreak system on the south coast of Western Australia. 1. Microclimate and wind erosion. *Australian Journal of Experimental Agriculture*. 42: 703-715.

Sudmeyer, R.A. and J. Speijers. 2007. Influence off windbreak orientation, shade and rainfall interception on wheat and lupin growth in the absence of below-ground competition. *Agroforestry Systems*. 71: 201-214.

Sudmeyer, R. A., M.A. Adams, J. Eastham, P.R. Scott, W. Hawkins and I.C. Rowland. 2002. Broadacre crop yield in the lee of windbreaks in the medium and low rainfall areas of south-western Australia. *Australian Journal of Experimental Agriculture*. 42: 739-750.

Vigiak, O., G. Sterk, A. Warren and L.J. Hagen. 2003. Spatial modelling of wind speed around windbreaks. *Catena*. 52: 273-288.

Yunusa, I.A.M., G.W. Brown, R.M. Kwong, G.R. Ronnfeldt, A. Crouch and M. Unkovich M., 2002. Integrating biodiversity and productivity on intensive farms: a potential role for shelter-belts in the Victorian Riverina. Agriculture for the Australian Environment. (http://live.greeningaustralia.org.au/nativevegetation/pages/pdf/Authors%20Y/2_Yunuasa_et_al.pdf).

In: Bioenergy Systems, Biological Sources
Editor: Michel C. Allard
ISBN: 978-1-62417-330-1

Chapter 6

Hydrogen and Ethanol Comparison as Fuels for Internal Combustion Engines

Verónica Laura Martínez and Juan Isidro Franco*
Instituto de Investigaciones Científicas y Técnicas para la Defensa
Buenos Aires, Argentina

Abstract

Bioethanol is one of the most important alternative fuels nowadays. Brazil and USA are the two main producers. In Brazil, it is produced from sugar cane and used in modified internal combustion engines, in vehicles called "flexfuel cars". Biohydrogen can also be produced from sugar (sucrose) and used as a fuel in a modified internal combustion engine. The advantages for hydrogen are zero carbon emissions and the possibility to use as a bridging technology for the introduction of proton exchange membrane fuel cell (PEMFC) vehicles. An analysis based on energy content, emissions, and production yield from sucrose, for each fuel, is presented here, along with the case of Argentina, where compressed natural gas (a gaseous fuel) is commonly used in the mass transportation system.

PACS 05.45-a, 52.35.Mw, 96.50.Fm

Keywords: Hydrogen as a fuel, dark fermentation, bioethanol, biohydrogen, compressed natural gas, internal combustion engine

AMS Subject Classification: 53D, 37C, 65P

1. Introduction

In the present day, the most widespread energy sources around the world come from contaminating and non renewable materials: fossil fuels, such as oil and gas, and coal (Figure 1). These non renewable sources will exhaust at some point and this fact, added to demand increase, makes the prices go up. Also, environmental pollution and greenhouse gases generated by their combustion are causing alternative energy sources (renewable and

*E-mail address: vmartinez@citedef.gob.ar

non -or less- polluting) to gain ground in recent years. This problems are specially important for the transportation sector. About 98% of the energy to power vehicles around the world comes from petroleum [1]. Among the alternative sources we find hydroelectric energy, wave and tidal energy from oceans, nuclear, solar, wind and biomass [2–8].

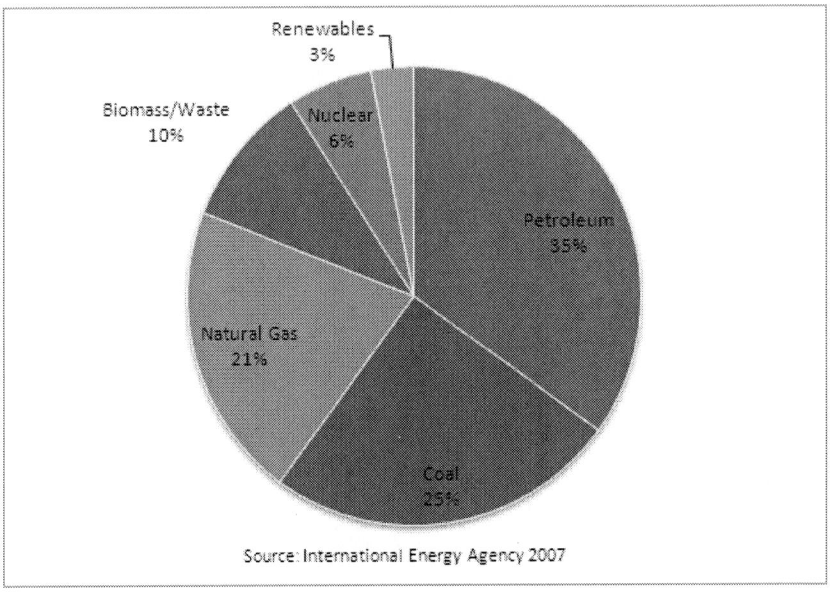

Figure 1. Energy sources around the world.

This last source – biomass – refers to organic matter. It is very attractive because it is renewable and it is possible to obtain from many different organic materials, such as waste from plant and tree crops, agricultural or agro-industrial waste, urban solid waste (USW) or specifically grown crops (called *energy crops*). Alternative methods to obtain energy, such as fermentation, are preferred over the traditional ones which mostly comprise fuel wood and burning to obtain heat. Not only does biomass have low heat content if burned, but it also creates polluting gasses and particles [9]. There exist several alternative fuels that are candidates to slowly replace fossil fuels for transportation. They are biodiesel, bioethanol and hydrogen.

Argentina has a huge energy production potential from these alternative sources since this country has a great extension of arable land and agricultural activities, strong Patagonian winds in the south, high levels of solar radiation in the North West and Central West regions (called *Noroeste* and *Cuyo*), important tidal differences and long coastal line thanks to the vast Southwestern Atlantic oceanic shelf [10–14].

In the last few years, research and development in this sector has received special attention because of its great potential. Many advances have been made. New technologies were developed such as generators, crop varieties, tillage and sowing techniques, machinery; and installed capacity has increased [15]. This major boost for the sector has also reached biofuels. Biodiesel and bioethanol are the two major biofuels in the country. Both can be obtained from renewable sources and, according to Anschau et al., emit reduced amounts of greenhouse gasses compared to fossil fuels [16].

Hydrogen is an energy vector. It is called so because it allows to transfer energy in space or in time and makes it available wherever and whenever needed [17]. It is included within the *alternative energy* denomination. It is very attractive because it is clean and it can be obtained from many different sources. If it is utilized in a PEM type fuel cell (proton exchange membrane) it does not produce contaminating emissions, but only water vapour [18]. When hydrogen is obtained through living organisms it is called biohydrogen and it falls under the category of *biofuels*. Both of these technologies, the PEM fuel cell and the production of biohydrogen, were recently developed in the research laboratories belonging to CITEDEF, the Institute of Scientific and Technological Research for Defense in Argentina.

2. Bioethanol

Bioethanol is fuel alcohol; or ethanol produced using biological methods, to use as a fuel in internal combustion engines. The main worldwide producers of ethanol as a fuel concentrate in North and South America, being the United States and Brazil the most important ones (Figure 2). Ethanol can also be produced by synthetic means, but according to the US Department of Energy [19], bioethanol represents over 90% of all ethanol production.

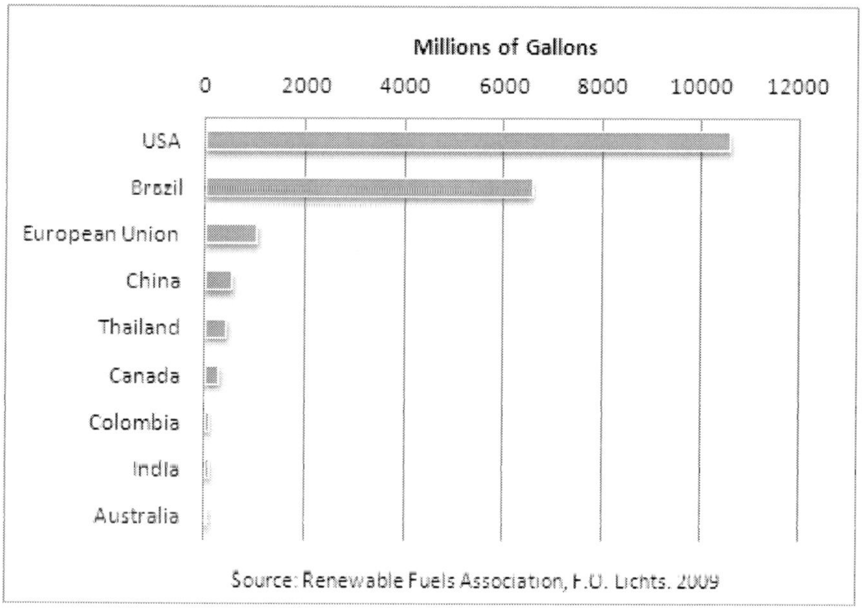

Figure 2. Ethanol production by country, in millions of gallons.

Fuel ethanol is produced in Brazil mainly from sugar cane and in the USA from corn. Fermentation is the chosen method to produce it, using microorganisms and sugar as their food. Since plants have sugar in their composition, any plant in theory could be used. However, there are crops that have higher concentration of sugar and processes that have been optimized for them.

2.1. Properties and Production

Ethanol is a liquid at ambient temperatures. It is an organic molecule with the formula C_2H_5OH, with a molecular weight of 46.07 g/mol, and a boiling point of 78.5^{o}C. The fact that it is a liquid makes it similar to gasoline when it comes to handling, storing and selling it in the mass market.

Gasoline has an octane rating of 88, while ethanol rating is 99. This means that an engine working on ethanol shows a higher compression ratio and higher power output per cycle. Ethanol flashpoint is 12.7^{o}C while that of gasoline is -43°C. This is the lowest temperature at which the liquid will ignite, therefore gasoline is highly flammable, but both are considered hazards since they ignite at ambient temperature. However, autoignition temperatures are above 250°C for both fuels (423°C for ethanol and 257°C for gasoline) which makes them rather stable in everyday life. Of course, certain precautions need to be taken into account all the same, such as lack of ignition sources in the vicinity of fuels' vapours [19].

Bioethanol from sugar cane is produced through fermentation followed by distillation. First the juice is squeezed from the stalks, and enters the fermentation process in a bioreactor. The dry stalks (called bagasse) can be burned to obtain energy and also use it in the production process. Ethanol fermentation is a metabolic pathway found in many different microorganisms. They posses the key enzyme pyruvate decarboxylase, in charge of decarboxylating pyruvate to acetaldehyde which later becomes ethanol. The most extensively used microorganism is the yeast belonging to the species *Saccharomyces cerevisiae*. This organism belongs to the Eukaryotes. Glucose enters glycolysis (Embden-Meyerhof pathway) where it is cleaved into two molecules of pyruvic acid. These, then, proceed to the alcohol fermentation pathway where they are converted to CO_2 and acetaldehyde which is in turn reduced to ethanol (Figure 3) [20].

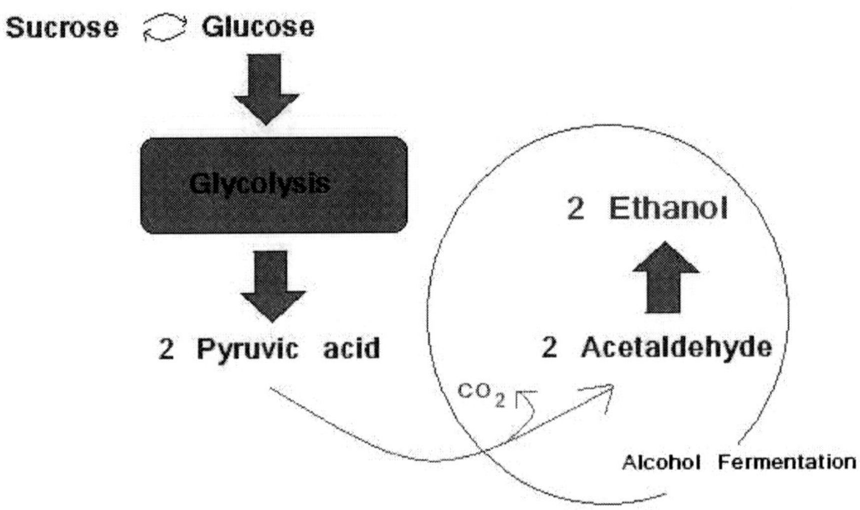

Figure 3. Glycolysis and alcohol fermentation.

Table 1. Comparison of production costs for bioethanol, in Euros/m^3, values from 2004 [24]

.	USA (maize)	Germany (beet)	Brazil (cane)
Production cost	394.7	595.7	144.8
Sale of byproducts	-67.1	-72.0	-
Gov. Subsidies	-79.3	-	-
Net cost	248.3	523.7	144.8

After the fermentation, the effluent from the reactor (called wine) undergoes a distillation process in columns to concentrate the ethanol. The water and impurities from the bottom of the columns (called vinasse) can be used for irrigation. The ethanol is then sieved as a final step [1]. Bioethanol produced from crops that could be used for human consumption is called *first generation*. The most widely used sugar-rich crops fed to the yeast are sugar cane (*Saccharum sp.*), sugar beet (*Beta vulgaris*), corn (*Zea mays*) and wheat (*Triticum* sp.). These are used due to their high concentration of readily available carbohydrates in the form of sucrose. Starchy crops require a pretreatment to increase glucose availability, and this of course, consumes energy. Sugar cane is the most productive, since it yields 8 times more energy than what the production process consumes (compare to corn which produces 1.34 times) [21]. In recent years, research has been intensely developed to find other plants suitable for the yeast but which are not destined to humans, such as *Miscanthus*. Bioethanol from these sources is called *second generation* [22].

Sugar cane juice and molasses are predominantly used in Brazil, where there is a extensive sugar cane production and harvest that spans over a period of 200 days throughout the year. The juice can be used to produce either crystalline sugar or ethanol. The production plants are typically as big as 500 million litres. Temperature requirements range from 30 to 35°C. The ethanol content of the effluent ranges from 6 to 11% v/v. After the distillation and sieving, the anhydrous alcohol has a purity of 99.9% v/v.

The average sugar cane yield amounts to 77,000 kg of sugar cane/ha. The stoichiometry of the ethanol conversion formula is as follows:

$$C_6H_{12}O_6 \longrightarrow 2\,C_2H_5OH + 2\,CO_2$$

To produce 1 litre of ethanol approximately 12 to 14 kg of sugar cane are needed. This represents around 6000 L/ha. All these values mentioned have been calculated for Brazil [23]. Molar masses are 342.3 g/mol of sucrose and 46.07 g/mol of ethanol. Knowing that ethanol density is 780 g/l, the yield is 0.48 mol ethanol per mol of sucrose.

As mentioned before, the sugar cane is highly efficient. It yields 8 units of energy for each unit consumed during the production process [21] [25]. In countries with extended sugar cane production, usually where the soil and weather are appropriate such as Brazil, production of bioethanol can be sustained without detriment to the food industry and at competitive prices, as shown in Table 1, constructed with values from the year 2004 [24] [26]. It can be seen how convenient the Brazilian production is, compared to USA where it is subsided by the government, or Germany where it is expensive.

Table 2. Comparison of fuel economy for different engines running on gasoline or ethanol (when possible) according to ADVISOR software models combining city and highway driving cycles [27]

Engine type	Gasoline (km/l)	Ethanol (km/l)
1.4 flexfuel	14.5	9.9
1.6 flexfuel	14.2	9.1
1.8 flexfuel	12.5	8.5
2.0 flexfuel	12.5	9.0
2.3 gasoline	11.4	-
2.4 flexfuel	10.4	7.1

2.2. Bioethanol as a Fuel

Nowadays, bioethanol and gasoline mixtures can be bought in the market. A blend of gasoline with a 5 to 20% of bioethanol can be used in a regular car (or other vehicle) with a regular internal combustion engine. Gasoline with 10% ethanol is termed "E10". Pure ethanol, or a blend of up to 15% of gasoline, can be used in modified combustion engines. The mix termed "E85" in the market has 85% ethanol and can be used in a type of vehicle commercially available in the present, called "flexfuel" car, for example Dodge Grand Caravan E85, Chevy Caprice E85, Buick Lacrosse 2012, Cadillac SRX E85, Chrysler E85 Flex Fuel, among others.

A study by W. F. Schmitt et al. [27] from the Petrobras Research Center and the Federal University of Rio de Janeiro, compares car efficiencies and fuel economy for gasoline and ethanol. The energy efficiencies were modelled by the ADVISOR software which requires well adjusted fuel consumption profiles of the internal combustion engine. According to several parameters for the vehicles, such as engine type, vehicle mass, volumetric density, and others, the driving cycles of an average Brazilian subcompact car were modelled with the software and validated simulating the cycles on a chassis dynamometer. Table 2 compares the fuel economy results with gasoline and ethanol for several engines.

Ethanol energy content is 37% lower than that of gasoline. This means that a car needs more ethanol per kilometre, as is shown in Table 2. But based on values available in 2005, the gasoline production cost was 0.59 \$/kg while bioethanol production costs were 0.30 \$/kg [28]. As a whole, driving with bioethanol is cheaper than driving with gasoline due to the low production cost from sugarcane. In 2005, the cost of driving 1 km with gasoline was 0.0393\$ /km, while driving 1 km on pure ethanol costed 0.0294\$/km.

The combustion product of bioethanol is CO_2, which is a greenhouse gas. Alcohol produces less CO_2 per **unit of mass** than fossil fuels. If we take, for instance, gasoline (C_8H_{18}, octane) the combustion reaction is as follows:

$$C_8H_{18} + 8.5\,O_2 \longrightarrow 8\,CO_2 + 9\,H_2O$$

While ethanol combustion reaction yields the following products:

$$C_2H_5OH + 3\,O_2 \longrightarrow 2\,CO_2 + 3\,H_2O$$

The molar mass of octane is 114 g/mol, and for ethanol is 46 g/mol. For CO_2 the molar mass is 44 g/mol which means that a kilogram of bioethanol releases into the atmosphere 1.91 kg of CO_2, while a kilogram of gasoline releases 3.09 kg of CO_2. This translates into the fact that gasoline produces 61.4% more greenhouse gas emissions than bioethanol per unit of mass.

Taking the values from Table 2 we see that a flexfuel car with a 1.4 engine can run 14.5 km with one litre of gasoline, but with a litre of ethanol it can only run 9.9 km. If we calculate the ethanol we need to run the 14.5 km, we find out the answer is 1.46 litres. This is expected since ethanol energy content is lower. Gasoline density is 0.68 kg/l, and for ethanol it is 0.78 kg/l. Therefore, to run the same distance the car consumes 1 litre of gasoline which equals to 0.68 g, while it needs 1.46 litres of ethanol which equals to 1.14 g. The CO_2 emissions for the felxfuel car when considering the distance covered by the drive (in this case 14.5 km) are 2.10 g of CO_2 with gasoline, and 2.18 g of CO_2 with ethanol. Theoretically, when considering fuels on a mole or litre basis, the greenhouse gas emissions are considerably less per mole or litre of ethanol. However, if the distance covered by the car is considered, the CO_2 emissions per kilometre are very similar, as was tested in current gasoline and flexfuel cars available in Brazil.

Since bioethanol comes from plant biomass, the same plants later take up this CO_2 from the vehicles and fix it again as a part of the carbon natural cycle. In this sense, engines running on ethanol are better than the ones running only on fossil fuels since the former do not contribute to the net increase of greenhouse gasses in the atmosphere. This is true as long as the plant biomass of the world uptakes the CO_2 at the same rate or faster than the vehicles produce it. Since the CO_2 from vehicles accumulates in cities, a possible way to palliate CO_2 accumulation is to design proper urban forestry programs to notoriously increase CO_2 uptake.

Ethanol has health benefits. Commercial gasoline includes many toxic additives, such as methyl tertiary-butyl ether or lead, which bioethanol does not contain. However, ethanol is not completely harmless. Its combustion produces a carcinogenic residue (formaldehyde) [29].

The advantages of bioethanol as a fuel are many, starting with the reduction of air pollutants and toxic emissions from vehicles operating on fossil fuels, replacement of non-renewable sources, diversification of energy sources and creation of jobs [26].

3. Biohydrogen

Biohydrogen ($bioH_2$) refers to hydrogen gas (H_2) produced through biological methods, that is, using microorganisms. In certain bacterial metabolic pathways H_2 is discarded as a by-product by the organism. It is generally mixed with other discarded gasses, mostly CO_2. Once purified, this biohydrogen is the same as any hydrogen gas produced through other methods, such as steam reforming or electrolysis.

H_2 is usually seen as a very desirable solution for the world's energy problems. A future energy system for the world based on H_2, known as the "hydrogen economy", can be envisioned. Ideally, this gas is abundant, non-polluting, can be obtained from water and other sources like biomass and every country would be able to produce the amounts they need. However, as ideal as it sounds, it is not yet operational since many difficulties have been encountered. Some technological barriers must be overcome before it can be produced, stored, transported and used in the mass market [30].

H_2 is not an energy source since no reserves of this gas exist on Earth. It is an energy carrier that needs to be produced. H_2 production is still expensive and derived from fossil fuels. For H_2 to be truly clean, it should be obtained from renewable energy sources derived from the sun, such as solar, biomass or wind. It should also be used in a PEM fuel cell (polymer electrolyte membrane fuel cell), a device that very efficiently converts H_2 energy into electricity producing only water vapour. The technology of PEM fuel cells for the mass market is under development.

Being the smallest gas of all, it is difficult to store and transport. It can be compressed up to several thousand psi in tanks, or even liquefied, but the compression can take up to one third of the stored energy, making it highly inefficient. Also, the tank metallic materials undergo a degradation process called "embrittlement" [31]. It can also be stored as a solid, in metal hydride or nano structured materials. This seems to be the best solution, but a better understanding of the atomic interactions between the gas and the material during H_2 absorption and release, and development of new materials are still needed.

Nowadays, it is not used as a fuel or in any application for the masses. The most widespread use is for the industry, such as synthesis of ammonia, and is derived from fossil fuels [32]. But many research institutions around the world are focusing their efforts on these particular barriers to finally develop reliable processes and devices to make the hydrogen economy a reality.

3.1. Properties and Production

Hydrogen is a gas at ambient temperature. Its molecular size is very small. Its boiling point is -252.8^{o}C. It is odourless, colorless, tasteless and burns with an invisible flame. Due to its small size it can easily leak out of containers. Its high diffusion velocity in air disperses it quickly, which makes it very unlikely to accumulate up to explosion limits in open spaces. The MSDS flammability index equals 4. When the right precautions are acknowledged and followed, H_2 is no more dangerous than propane gas in households (which also has an index of 4). The energy yield of hydrogen (combustion enthalpy variation per gram) is 122 kJ/g, which is 2.75 times greater than that of hydrocarbon fuel [33].

Hydrogen gas can be produced using fossil fuels and certain procedures that release the hydrogen contained in the hydrocarbon molecules. These processes consume energy and release contaminants and greenhouse gasses into the atmosphere. Traditional processes include natural gas steam reforming, coal and biomass gasification and water electrolysis [34] [35]. To sustain these processes a CO_2 capture system is needed. The most widely used are post-combustion capture, pre-combustion capture, oxy-combustion and chemical looping. After being captured, the CO_2 has to be stored for long periods of time, for example, in exhausted oil and gas reservoirs [36]. All these complications derived from the usage of fossil fuels make other sources of H_2 very desirable.

In recent years, biological production of hydrogen has received quite a lot of attention from the scientific community around the world. Several microorganisms are capable of producing H_2 through several metabolic pathways. They can be cultured in big tanks or reactors under very controlled conditions to optimize H_2 yield. Production can be coupled to water treatment, waste degradation or CO_2 capture, depending on which process is employed.

The first of these processes is called "bio-photolysis" of water. It happens due to the action of the hydrogenase enzyme at the end of another very well known process, the photosynthesis [37]. Some algae, for example *Chlamydomonas reinhardtii*, which are eukaryotic cells possess the enzyme and are widely used for research in this area. Some cyanobacteria, for example *Anabaena* sp., which are prokaryotic cells also produce H_2 because they also have the enzyme. They need light, CO_2 and anaerobic conditions, since the enzyme is inhibited by O_2 [38] [39]. These photosynthetic organisms remove CO_2 from the atmosphere and use it to produce biomass as they grow. However, more research is still needed to improve the hydrogen yields [37] [40] [41].

The second metabolic process that produces H_2 is "photo-fermentation". The H_2 production here is linked with the activity of the nitrogenase enzyme in purple non-sulphur bacteria, such as *Rhodobacter sphaeroides*, *Rhodopseudomonas capsulatus* or *Rhodospirillum rubrum*. These bacteria can use small organic acids as carbon source, which makes it possible to couple this reactor tank to the effluents from the tank where the next process occurs (dark fermentation); they can be set up in a sequence of fermentations to improve H_2 yield [37] [42].

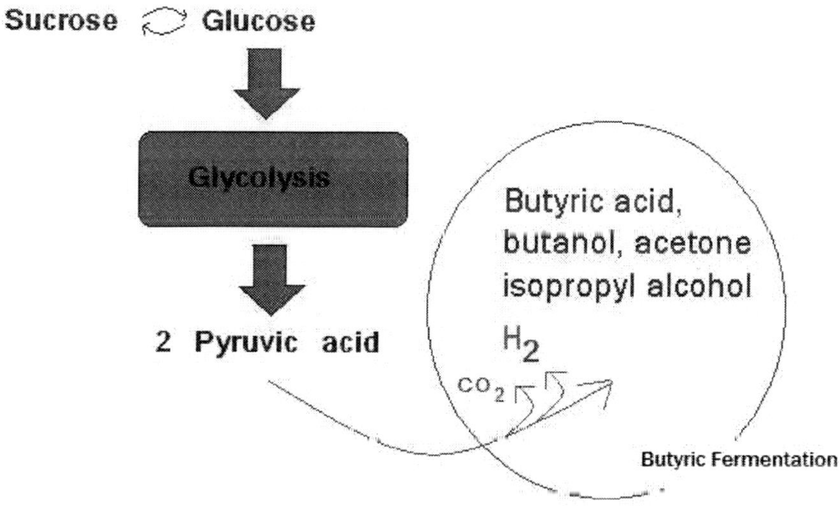

Figure 4. Glycolysis and butyric fermentation.

The third interesting pathway is "dark fermentation", which is performed by certain anaerobic bacteria that play a part in the anaerobic digestion process of decaying organic matter and wastes. These species, such as *Clostridium* sp. and *Enterobacter* sp. among a few others, take up carbohydrates from the exterior (from sucrose or biomass) and obtain energy through the glycolysis pathway, exactly like the yeast *Saccharomyces cerevisiae* does, shown in the previous section for bioethanol. However, instead of alcohol fermentation, enterobacteria feature mixed acid fermentation and produce ethanol, CO_2, bioH_2 and a few carboxylic acids such as acetic, butyric, propionic, etcetera [43]. Clostridia metabolize through butyric acid fermentation, therefore they produce bioH_2, CO_2, acetic acid and butyric acid, see Figure 4 [18]. This small-chain organic acids are the substrates for the purple non-sulphur bacteria mentioned in the paragraph above. For an in depth description of the anaerobic digestion process, please refer to the following section of this publication "*Hydrogen Production by Mixed Cultures*".

This last process – dark fermentation – is very convenient, financially viable and is subject of much research. Different and easily available bacterial communities have been tested in our laboratory for H_2 production, such as sludge, dirt or compost [44]. These communities have been cultured in batch reactors with sucrose. Culture media and bacteria are incubated until the sucrose is consumed and H_2 production is assessed at different times during the batch. It is also possible to culture them in a continuous reactor, where exhausted culture media is removed and replaced with fresh media with sucrose and nutrients. In this way, H_2 production can be sustained throughout a long period of time.

The bacterial metabolism of dark fermentation, according to the products, is as follows (for more details refer to the next chapter of this publication "*Hydrogen Production by Mixed Cultures*"):

$$C_6H_{12}O_6 + 2\,H_2O \longrightarrow 2\,CH_3COOH + 2\,CO_2 + 4\,H_2$$
$$C_6H_{12}O_6 \longrightarrow CH_3(CH_2)_2COOH + 2\,CO_2 + 2\,H_2$$

As shown, the acetate pathway yields 4 moles of H_2, while the butyrate pathway yields 2. In a bacterial community, both pathways occur at the same time; depending on the pH and pressure conditions one might predominate over the other. Therefore, the theoretical maximum yield of 4 moles of H_2 per mole of glucose is never achieved. The experimental yield in some continuous reactors that have been operational for months were 1.04, 1.74, 2.25 or as high as 2.35 moles of H_2 per mole of glucose [45] [46] [47] [48]. All of these molar yields are highly superior to the bioethanol molar yields obtained with sugarcane (see above, Bioethanol Properties and Production section).

This great variety, flexibility and yield makes biohydrogen a convenient option. Biomass as substrate means that many agro-industrial residues or even solid urban waste have potential to be a hydrogen source. CO_2 fixation in the case of biophotolysis has the potential to reduce greenhouse gasses in the atmosphere. Both these features are great advantages of biohydrogen over the other traditional hydrogen production methods.

3.2. Biohydrogen as a Fuel

Fuel cells (FC) are devices that convert chemical energy to electricity. They are not subject to Carnot cycle efficiencies, as combustion engines are, and this fact makes them notably more efficient than the latter. A combustion engine is about 20-30% efficient, while a fuel cell is 75-80% efficient (without taking into account the energy consumed in the production of the fuel) [21].

The most common fuel for the fuel cells is hydrogen and the most efficient type of cell is the solid oxide FC. However, they operate at very high temperatures. One very convenient type of cell for portable applications that operates at low temperature is the polymer electrolyte membrane, or PEM (also proton exchange membrane) [21]. This type of FC can be used to power cars and other means of mass transportation.

Hydrogen and PEM fuel cells are a very attractive alternative to power vehicles. Emissions are truly *green* since the only product of a PEMFC is water vapour and the device is highly efficient, using about 80% of the energy comprised in the H_2 molecule. These are important reasons why H_2 for fuel cells is a very interesting fuel in development. Vehicles powered by hydrogen fuel cells were termed H_2FC [49].

Hydrogen gas can also be used in an internal combustion engine. The engine needs a few modifications before it can use this gas, mainly due to the physical properties of H_2: high diffusion velocity in air, flammability, low ignition energy, low energy content per unit of volume. In May 2000, the German auto mobile manufacturer BMW presented the "BMW 750 hL" model that works on both gasoline and hydrogen. This 12-cylinder sedan built with all the comfort of a BMW can reach 226 km/h turning polluting emissions to virtually zero, since the results of hydrogen combustion is water vapour [50]. To avoid self ignition and prevent polluting NO_x emissions a lean mixture with air must be used ($\gamma \geq 1.8$). To avoid backfiring during the intake stroke, hydrogen accumulation in the intake manifold must be eliminated. This is achieved by sequential multi-point injection for each cylinder at the intake stroke [51]. A regular Otto engine from any commercial car can be adjusted to these requirements as shown in a Ford V-8 engine with a Holley carburettor [52] and a Ford Escort [51].

4. Biohydrogen vs. Bioethanol for Otto Engines

Hydrogen gas has a lower energy content than ethanol and gasoline. The combustion enthalpy for octane is -5.33 MJ/mol, while that of ethanol is -1370.7 kJ/mol and hydrogen is -286 kJ/mol. However, when we compare the similar production methods for biohydrogen and bioethanol it is evident that the yield of the fermentation process is better for biohydrogen with an average of 1.74 mol H_2/mol glucose, versus ethanol with 0.48 mol EtOH/mol sucrose. This would mean that for each mole of carbohydrate fed into the fermenter an average of -658 kJ in bioethanol could be obtain, while -503.4 kJ in biohydrogen are obtained. In the case of the highest yields obtained for biohydrogen fermentations, 4.70 mol H_2/mol sucrose [48], in terms of energy the yield is as high as -1344.2 kJ/mol sucrose.

Using hydrogen in internal combustion engine (ICE) cars is possible. The engine needs a few simple modifications to be able to run on the combustion of a gaseous fuel. These vehicles were termed H_2ICE [49]. Using hydrogen in cars would eliminate toxic emissions completely, while bioethanol combustion still produces CO_2 in big quantities. Also, modifying a car to run on hydrogen is easier and cheaper than for ethanol. The advantages for ethanol are the similarities it shares with gasoline. Hydrogen, on the other hand, is a gas at ambient temperature. Its liquefaction is extremely energy consuming which makes the fuel inefficient. It can be stored as compressed gas in tanks, but they occupy much car space. Hydrogen storage technologies are expected to improve in the near future, especially solid storage materials.

Fuelling stations should also be adapted to dispatch gas instead of liquid fuel. This barrier can be overcome, as is shown in the case of Argentina where natural compressed gas is commonly used as a fuel for regular vehicles and sold without inconveniences in the mass market (see next section).

5. Biohydrogen and Natural Gas Mixture as a Fuel

Natural gas (NG) is an alternative to gasoline commercially used in the present day. Natural gas is composed mostly of methane and ethane, and it is a by-product of crude oil. It is often found dissolved in the oil in a reservoir, or as a gas cap above the oil. It can be compressed in tanks, known as CNG, and used as a fossil fuel to power ICE vehicles. The combustion of CNG is more efficient than gasoline, however air pollutants are released anyway, such as carbon monoxide (CO) and unused hydrocarbons (HC), though, in less quantities than gasoline [53]. A drawback for CNG as a fuel is similar to that of hydrogen: a high quantity of the fuel is needed due to their the low energy density, but since it is a gas, it takes up too much storage space in the vehicle. Fuelling stations need gas pipes and pumps.

In Argentina, CNG is cheaper than gasoline, which drove many car owners to convert their car to a gas ICE vehicle. The yellow gas tank is quite big and it is kept in the trunk. Many fuelling stations have gas infrastructure. The National Regulatory Entity for NCG (ENARGAS) published 2009 statistics registering a total of 1827 fuelling stations and over 2 million NCG vehicles in the country [54]. This infrastructure and public acceptance of a gas fuel make it a perfect location for an introduction of hydrogen as a fuel for ICE vehicles. In the present day, there is a project aimed at producing a mixture of CNG and 20% of H_2. This mix lowers CO, CO_2 and HC emissions, but increases NO_x emissions [55]. When taking into account the production process of hydrogen from fossil fuels, life cycle analysis for the NCG-H_2 mix results in an increase of greenhouse gas emissions [56]. Production form biomass, sucrose or hydrolysis through renewable energy sources such as solar or wind, turns this fuel into a very promising alternative. These cars could also act as a bridging technology between the current ICE vehicles powered with gasoline and future fuel cell vehicles powered with hydrogen.

References

[1] Cesar B. Granda, Li Zhu, and Mark T. Holtzapple. Sustainable liquid biofuels and their environmental impact. *Environmental Progress*, 26(3):233–250, 2007.

[2] Geoffrey P. Sims. Hydroelectric energy. *Energy Policy*, 19(8):776 – 786, 1991.

[3] A. Al-Habaibeh, D. Su, J. McCague, and A. Knight. An innovative approach for energy generation from waves. *Energy Conversion and Management*, 51(8):1664 – 1668, 2010. Global Conference on Renewables and Energy Efficiency for Desert Regions (GCREEDER 2009).

[4] Fergal O'Rourke, Fergal Boyle, and Anthony Reynolds. Tidal energy update 2009. *Applied Energy*, 87(2):398 – 409, 2010.

[5] Manfred Lenzen. Life cycle energy and greenhouse gas emissions of nuclear energy: A review. *Energy Conversion and Management*, 49(8):2178 – 2199, 2008.

[6] Nelson A. Kelly and Thomas L. Gibson. Increasing the solar photovoltaic energy capture on sunny and cloudy days. *Solar Energy*, 85(1):111 – 125, 2011.

[7] Ezio Sesto and Claudio Casale. Exploitation of wind as an energy source to meet the world's electricity demand. *Journal of Wind Engineering and Industrial Aerodynamics*, 74-76(0):375–387, 1998.

[8] M. Hoogwijk, A. Faaij, R. van den Broek, G. Berndes, D. Gielen, and W. Turkenburg. Exploration of the ranges of the global potential of biomass for energy. *Biomass and Bioenergy*, 25(2):119 – 133, 2003.

[9] Mustafa Balat. Use of biomass sources for energy in Turkey and a view to biomass potential. *Biomass and Bioenergy*, 29(1):32 – 41, 2005.

[10] The World Bank Group. Climate change knowledge portal. http://sdwebx.worldbank.org/climateportal/index.cfm?page=country_impacts_agriculture&ThisRegion=Latin%20America&ThisCcode=ARG, 2012.

[11] M. Recalde. Wind power in Argentina: Policy instruments and economic feasibility. *International Journal of Hydrogen Energy*, 35(11):5908 – 5913, 2010. 3rd Argentinean and 2nd Latin American Congress in Hydrogen and Sustainable Energy Sources.

[12] Hugo Grossi Gallegos and Raúl Righini. Atlas de energía solar de la República Argentina. http://www.gersol.unlu.edu.ar/pagina3.htm, 2007. Universidad Nacional de Luján.

[13] Martin Saraceno, E.E. D'Onofrio, M.E. Fiore, and W.H. Grismeyer. Tide model comparison over the Southwestern Atlantic shelf. *Continental Shelf Research*, 30(17):1865 – 1875, 2010.

[14] Santiago & Sinclair LLC. The Clean Energy Report: Estado de la industria argentina de energías renovables. http://www.santiagosinclair.com/files/Estado_de_la_industria_argentina_de_energias_renovables.pdf, 2011.

[15] Jorge Schvarzer and Andrés Tavosnanska. Biocombustibles: expansión de una industria naciente y posibilidades para Argentina. http://www.cosechaypostcosecha.org/data/articulos/biocombustibles/Biocombustibles_ExpansionIndustriaNaciente.pdf, 2007. Documento N° 13, CESPA, FCE, Universidad de Buenos Aires.

[16] R. A. Anschau, N. Flores Marco, S.M. Carballo, and J. Hilbert. Evaluación del potencial de producción de biocombustibles en Argentina. http://www.cpa.unicamp.br/alcscens/publications/Evaluacion_Produccion_biocombustibles.pdf, 2009. Instituto de Clima y Agua e Instituto de Ingeniería Rural, INTA.

[17] Fabio Orecchini and Vincenzo Naso. *Energy Systems in the Era of Energy Vectors*. Green Energy and Technology. Springer, 2012.

[18] Verónica L. Martínez, Rodrigo E. García, Gustavo Curutchet, Alfredo Sanguinetti, Héctor J. Fasoli, and Juan I. Franco. Demonstration of the possibility to power a fuel cell with hydrogen derived from the fermentation of sugar. *International Journal of Hydrogen Energy*, (0):–, 2012.

[19] U.S. Department of Energy. Biomass Energy Data Book. http://cta.ornl.gov/bedb/biofuels.shtml, 2011.

[20] Gerard J. Tortora, Berdell R. Funke, and Christine L. Case. *Microbiology: An Introduction*. Pearson Benjamin Cummings, 2010.

[21] Liliana Hernández and Viatcheslav Kafarov. Use of bioethanol for sustainable electrical energy production. *International Journal of Hydrogen Energy*, 34(16):7041 – 7050, 2009. 4th Dubrovnik Conference.

[22] European Biomass Industry Association. Bioethanol. http://www.eubia.org/212.0.html, 2012.

[23] David Pimentel and Tad Patzek. Ethanol production: Energy and economic issues related to U.S. and Brazilian sugarcane. *Natural Resources Research*, 16:235–242, 2007.

[24] José Goldemberg and Patricia Guardabassi. Are biofuels a feasible option? *Energy Policy*, 37(1):10 – 14, 2009.

[25] Harro von Blottnitz and Mary Ann Curran. A review of assessments conducted on bioethanol as a transportation fuel from a net energy, greenhouse gas, and environmental life cycle perspective. *Journal of Cleaner Production*, 15(7):607 – 619, 2007.

[26] Suani Teixeira Coelho, José Goldemberg, Oswaldo Lucon, and Patricia Guardabassi. Brazilian sugarcane ethanol: lessons learned. *Energy for Sustainable Development*, 10(2):26 – 39, 2006.

[27] William F. Schmitt, Alexandre Szklo, and Roberto Schaeffer. Policies for improving the efficiency of the brazilian light-duty vehicle fleet and their implications for fuel use, greenhouse gas emissions and land use. *Energy Policy*, 39(6):3163 – 3176, 2011.

[28] Lin Luo, Ester van der Voet, and Gjalt Huppes. Life cycle assessment and life cycle costing of bioethanol from sugarcane in Brazil. *Renewable and Sustainable Energy Reviews*, 13(6-7):1613 – 1619, 2009.

[29] Graeme M. Walker. 125th anniversary review: Fuel alcohol: Current production and future challenges. *Journal of the Institute of Brewing*, 117(1):3–22, 2011.

[30] U.S. Department of Energy. Basic research needs for the hydrogen economy. http://science.energy.gov/~/media/bes/pdf/reports/files/nhe_rpt.pdf, May 13-15 2003. Basic Energy Sciences Workshop on Hydrogen Production, Storage, and Use. Rockville, Maryland.

[31] H. Barthélémy. Effects of pressure and purity on the hydrogen embrittlement of steels. *International Journal of Hydrogen Energy*, 36(3):2750 – 2758, 2011. The Third Annual International Conference on Hydrogen Safety.

[32] Miroslava Smitkova, František Janíček, and Juri Riccardi. Life cycle analysis of processes for hydrogen production. *International Journal of Hydrogen Energy*, 36(13):7844 – 7851, 2011. Hysydays.

[33] Sakchai Pattra, Suksaman Sangyoka, Mallika Boonmee, and Alissara Reungsang. Bio-hydrogen production from the fermentation of sugarcane bagasse hydrolysate by *Clostridium butyricum*. *International Journal of Hydrogen Energy*, 33(19):5256 – 5265, 2008. 2nd Asian Bio Hydrogen Symposium.

[34] F. Mueller-Langer, E. Tzimas, M. Kaltschmitt, and S. Peteves. Techno-economic assessment of hydrogen production processes for the hydrogen economy for the short and medium term. *International Journal of Hydrogen Energy*, 32(16):3797 – 3810, 2007. TMS06: Symposium on Materials in Clean Power Systems.

[35] Calin-Cristian Cormos. Hydrogen production from fossil fuels with carbon capture and storage based on chemical looping systems. *International Journal of Hydrogen Energy*, 36(10):5960 – 5971, 2011.

[36] Working Group III of the Intergovernmental Panel on Climate Change [B. Metz, O. Davidson, H. C. de Coninck, M. Loos, and L. A. Meyer (eds.)]. IPCC special report on carbon dioxide capture and storage. Technical report, Intergovernmental Panel on Climate Change, 2005.

[37] Ilgi Karapinar Kapdan and Fikret Kargi. Bio-hydrogen production from waste materials. *Enzyme and Microbial Technology*, 38(5):569 – 582, 2006.

[38] Krishnan Vijayaraghavan, Rajendran Karthik, and S.P. Kamala Nalini. Hydrogen production by *Chlamydomonas reinhardtii* under light driven sulfur deprived condition. *International Journal of Hydrogen Energy*, 34(19):7964 – 7970, 2009.

[39] Sergei A. Markov, Michael J. Bazin, and David O. Hall. Hydrogen photoproduction and carbon dioxide uptake by immobilized anabaena variabilis in a hollow-fiber photobioreactor. *Enzyme and Microbial Technology*, 17(4):306 – 310, 1995.

[40] Tatyana Laurinavichene, Irena Tolstygina, and Anatoly Tsygankov. The effect of light intensity on hydrogen production by sulfur-deprived *Chlamydomonas reinhardtii*. *Journal of Biotechnology*, 114(1-2):143 – 151, 2004.

[41] Khorcheska A. Batyrova, Anatoly A. Tsygankov, and Sergey N. Kosourov. Sustained hydrogen photoproduction by phosphorus-deprived *Chlamydomonas reinhardtii* cultures. *International Journal of Hydrogen Energy*, 37(10):8834 – 8839, 2012. International Conference: Photosynthesis Research for Sustainability.

[42] Türker Lemi, Gümüs Selcuk, and Tapan Alper. Biohydrogen production: molecular aspects. *Journal of Scientific and Industrial Research*, 67, 2008.

[43] Andrea K. Forrest, Melinda E. Wales, and Mark T. Holtzapple. Thermodynamic prediction of hydrogen production from mixed-acid fermentations. *Bioresource Technology*, 102(20):9823 – 9826, 2011.

[44] Rodrigo Enrique García, Verónica Laura Martínez, Juan Isidro Franco, and Gustavo Curutchet. Selection of natural bacterial communities for the biological production of hydrogen. *International Journal of Hydrogen Energy*, 37(13):10095 – 10100,

2012. Proceedings of the Symposium on Hydrogen Production and Applications at the 240th American Chemical Society National Meeting, August 22-26, 2010, Boston, Massachusetts, USA.

[45] Chiu-Yue Lin, Shu-Yii Wu, Ping-Jei Lin, Jo-Shu Chang, Chun-Hsiung Hung, Kuo-Shing Lee, Chyi-How Lay, Chen-Yeon Chu, Chin-Hung Cheng, Alex C. Chang, Jou-Hsien Wu, Feng-Yuan Chang, Lee-Hao Yang, Chia-Wen Lee, and Yi-Chun Lin. A pilot-scale high-rate biohydrogen production system with mixed microflora. *International Journal of Hydrogen Energy*, 36(14):8758 – 8764, 2011. 4th Asian Bio-Hydrogen Symposium.

[46] I. Mariakakis, P. Bischoff, J. Krampe, C. Meyer, and H. Steinmetz. Effect of organic loading rate and solids retention time on microbial population during bio-hydrogen production by dark fermentation in large lab-scale. *International Journal of Hydrogen Energy*, 36(17):10690 – 10700, 2011. International Conference on Hydrogen Production (ICH2P)-2010.

[47] Kuo-Shing Lee, Tzu-Shan Tseng, Yen-Wen Liu, and Yi-Dai Hsiao. Enhancing the performance of dark fermentative hydrogen production using a reduced pressure fermentation strategy. *International Journal of Hydrogen Energy*, (0):–, 2012.

[48] Chin-Chao Chen, Hong-Pin Chen, Jou-Hsien Wu, and Chiu-Yue Lin. Fermentative hydrogen production at high sulfate concentration. *International Journal of Hydrogen Energy*, 33(5):1573 – 1578, 2008. <ce:title>Asian bioH2 Symposium 2006</ce:title>.

[49] Sebastian Verhelst and Thomas Wallner. Hydrogen-fueled internal combustion engines. *Progress in Energy and Combustion Science*, 35(6):490 – 527, 2009.

[50] BMW AG. The bmw hydrogen 7. `http://www.bmw.com/com/en/insights/technology/efficient_dynamics/phase_2/clean_energy/bmw_hydrogen_7.html`, 2012.

[51] Beckmann W., Bröcker J., Sponholz C., and Krüger M. Hydrogen as a fuel for internal combustion engines. `http://www.wti-mv.de/kompetenzen/beckmann_hydrogen.pdf`, 2012.

[52] Carlos Vidal, Gerardo Friedlmeier, Alejandro Yawny, and Juan Carlos Bolcich. Hacia un empleo masivo del hidrógeno como portador de energía. *Ingeniería Militar (upon request)*, 17(42):8–11, December 2000.

[53] Dušan Gruden. Fuels. In *Series Anthropogenic Compounds*, volume 3T of *The Handbook of Environmental Chemistry*, pages 255–288. Springer Berlin / Heidelberg, 2003. 10.1007/b10462.

[54] National Regulatory Entity for Compressed Natural Gas. GNC, el combustible limpio. http://www.gnc.org.ar, 2012.

[55] Instituto de Energía y Desarrollo Sustentable. Desarrollo de combustible híbrido gaseoso para medios de transporte público de pasajeros y de carga. http://www.cab.cnea.gov.ar/ieds/extras/actividades/gnc_h.pdf, 2012.

[56] P. Martínez, L. Dawidowski, D. Gómez, and D. Pasquevich. Life cycle greenhouse emissions of compressed natural gas-hydrogen mixtures for transportation in Argentina. *International Journal of Hydrogen Energy*, 35(11):5793 – 5798, 2010. 3rd Argentinean and 2nd Latin American Congress in Hydrogen and Sustainable Energy Sources.

In: Bioenergy Systems, Biological Sources
Editor: Michel C. Allard
ISBN: 978-1-62417-330-1

Chapter 7

HYDROGEN PRODUCTION BY MIXED CULTURES

Franco Juan Isidro and Rodrigo Enrique García*
Instituto de Investigaciones Científicas y Técnicas para la Defensa
Buenos Aires, Argentina

Abstract

In a process known as anaerobic digestion (AD) a consortia of microorganisms degrades organic matter to digested sludge, carbon dioxide and methane. During this process hydrogen is both produced and consumed as an important intermediary. Hence, if you wish to produce hydrogen through biological means, it becomes necessary to intervene in the anaerobic digestion process. In this short review we describe how is it possible to obtain hydrogen through the fermentation of carbohydrates, describing the metabolic pathways and clarifying the destiny and function of the several molecules involved. We hope that the information synthesized here provides clarity to the collection of information generally found in scientific publications.

PACS 05.45-a, 52.35.Mw, 96.50.Fm

Keywords: Hydrogen from mixed cultures, dark fermentation, biohydrogen, fermentation metabolism, anaerobic digestion

AMS Subject Classification: 53D, 37C, 65P

1. Introduction

Oil reserves exhaustion and the recognized environmental damage that the use of fossil fuels causes evidence the need to find an alternative fuel that does not harm the environment. As it is described by the United States Environmental Protection Agency (EPA) [1], alternative fuels include gaseous fuels such as hydrogen, natural gas and propane, alcohols such as ethanol, methanol and butane, oils derived from plant biomass and wastes. These fuels offer an alternative to gasoline and diesel. Similarly, the EPA defines renewable fuels as those derived from renewable sources not related to oil such as harvest, animal wastes or solid municipal wastes.

*E-mail address: rodrigo.egarcia@gmail.com

Hydrogen is an alternative, renewable fuel that is not harmful to the environment. Hydrogen combustion does not generate greenhouse gases or toxic gases such as carbon monoxide (CO), ozone (O_3) or sulphur oxides (SO_x) and only minimal amounts of Nitrogen oxides (NO_x) will be produced if air is used as oxidant [2]. In addition, hydrogen will not undergo spontaneous combustion or decay. It does not detonate on open environments. It will not act as an oxidant or degrade other compounds. In regards to human health, hydrogen is not toxic, corrosive, radioactive, teratogenic or carcinogenic [3]. All the reasons detailed above justify the research and development of non-conventional and environmentally friendly methods for the production of hydrogen.

Hydrogen gas can be produced in a renewable manner through biotechnology. Biohydrogen or biological hydrogen, as it is known when it originates from a renewable source [4], can be obtained from the anaerobic fermentation of carbohydrates by mixed microbial communities. There are many microorganisms and methods that can be used to produce hydrogen, the four most studied methods include direct biophotolysis, indirect biophotolysis, photo fermentation and dark fermentation. Each technique presents advantages and disadvantages when compared to the others and these should be carefully evaluated before choosing a method or a particular microorganism. In our laboratory at the Alternative Energy Research Department (DIDER), we have focused our research in the production of hydrogen through dark fermentation of simple carbohydrates (e.g. glucose and sucrose) with mixed cultures. The microbial communities used in our studies were obtained from sludge from a wastewater treatment plant located at one of our war ships.

We opted to produce hydrogen through dark fermentation because the required technology is simple and higher yields can be attained with this technique [4] than with the other methods. Likewise, this process does not depend on light and a large number of diverse substrates can be used. For example, mixed cultures will degrade complex and simple sugars, fats, proteins, biomass and wastes. Initially, it was our intention to ferment organic wastes, being the objective to produce a renewable fuel with simultaneous waste material degradation. However, as we executed a series of experiments [5] we found that the gas obtained when fermenting glucose and sucrose is a relatively clean gas that can be used without previous purification in Proton Exchange Membrane (PEM) fuel cells to produce electricity. Organic waste will not only contain carbohydrates but also proteins and fats, and therefore its fermentation will be associated with the production of gasses that are harmful to the PEM fuel cell such as hydrogen sulphide.

Furthermore, the diminished potential to produce hydrogen that protein-rich waste and fats-rich waste exhibit when compared to carbohydrate rich waste [4] and the fact that simple sugars are readily degraded by bacteria, justifies the use of these sugars as feedstock and not only as model substrates for research. In addition, the energy yield of hydrogen produced from sucrose with current technologies is competitive with the use of sucrose for bio-ethanol production. More energy is obtained per mol of sucrose when hydrogen is produced than when it is used to generate bio-ethanol as it was explained in the section, "<Hydrogen and Ethanol Comparison as Fuels for Internal Combustion Engines">.

In the next section we will describe how is it possible to obtain hydrogen through the fermentation of carbohydrates, describing the metabolic pathways and clarifying the destiny and function of the several molecules involved. We hope that the information synthesized provides clarity to the collection of information generally found in other scientific publications.

2. Carbohydrate Fermentation by Mixed Cultures

2.1. Anaerobic Digestion: The Big Picture

To comprehend how hydrogen is produced through the fermentation of carbohydrates by mixed cultures, first we need to understand a more general process that includes fermentation. This process is known as Anaerobic Digestion (AD), in AD different groups of bacteria degrade organic matter producing mainly digested sludge, carbon dioxide and methane. It is possible to identify four complementary and vital processes [6–8] in AD that have to take place in order to completely digest organic matter (Figure 1).

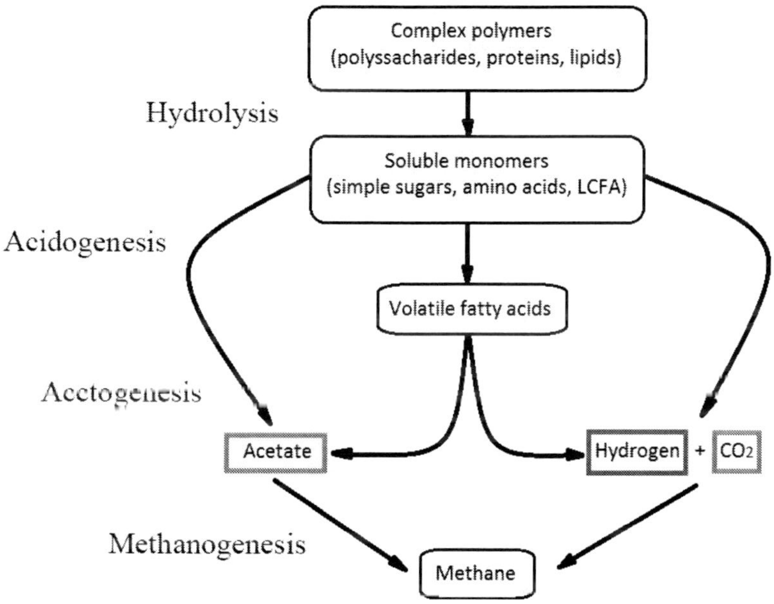

Figure 1. Modified from 7: Stages of anaerobic digestion (AD) of organic matter. Acetate is the preferred fermentation end-product from the stand point of hydrogen production.

The first one is the *hydrolysis* of the complex polymers to soluble monomers (equation 1). Polysaccharides are converted to simple carbohydrates, proteins to amino acids and lipids to long chain fatty acids (LCFA) by the action of hydrolytic enzymes released by bacteria. Once the polymeric material has been hydrolysed then *acidogenesis* and *acetogenesis* can occur. In acidogenesis, sugars and amino acids are degraded by fermentative bacteria to intermediary metabolites and hydrogen is formed (equation 2). Subsequently, in acetogenesis a new group of microorganisms called acetogenic bacteria will produce large quantities of acetate, carbon dioxide and hydrogen through the oxidation of LCFA and of other intermediary metabolites formed during hydrolysis and acidogenesis, such as propionate, butyrate, hydrogen and carbon dioxide (equations 3 to 5). Note that in acetogenesis hydrogen is both produced and consumed. The last process involved in AD, that can take place when products form acetogenesis are available, is *methanogenesis*. Here two new groups of bacteria will generate methane, water and carbon dioxide consuming acetate, hydrogen and alcohols (equations 6 and 7).

$$C_6H_{10}O_4 + 2H_2O \longrightarrow C_6H_{12}O_6 + 2H_2 \tag{1}$$

Hydrolysis: Organic waste broken down into glucose [7].

$$4C_6H_{12}O_6 + 2H_2O \longrightarrow 2CH_3COOH + 2CH_3(CH_2)_2COOH + 8CO_2 + 8H_2 \tag{2}$$

Acidogenesis: Glucose fermented to propionic and butyric acid by a mixed culture [9].

$$CH_3CH_2CH_2COOH + 4H_2O \longrightarrow CH_3COOH + 2CO_2 + 6H_2 \tag{3}$$

Acetogenesis: Butyric acid converted to acetic acid by obligate H_2-producing acetogenic bacteria [10].

$$CH_3CH_2COOH + 2H_2O \longrightarrow CH_3COOH + CO_2 + 3H_2 \tag{4}$$

Propionic acid converted to acetic acid obligate H_2-producing acetogenic bacteria [10].

$$4H_2 + 2CO_2 \longrightarrow CH_3COOH + 2H_2O \tag{5}$$

Homoaetogenic bacteria producing acetic acid from hydrogen and carbon dioxide, modified from [11].

$$CH_3COOH \longrightarrow CH_4 + CO_2 \tag{6}$$

Methanogenesis: Acetate conversion to methane by acetoclastic methanogen bacteria [10].

$$4H_2 + CO_2 \longrightarrow CH_4 + 2H_2O \tag{7}$$

Methane production by Hydrogenotrophic methanogens [10].

Due to the synergic interaction of these four processes the complete digestion of organic matter to methane, carbon dioxide and water, is achieved. In regards to the link between AD and the production of hydrogen, it must be stated that the end point of AD is the formation of methane. Hydrogen is generated as an intermediary and it will be consumed during methanogenesis. Also, some hydrogen will be lost to acetate formation reactions since some acetogenic bacteria (homoacetogens), have the capacity to grow consuming hydrogen and carbon dioxide via the Wood-Ljungdahl [11] metabolic pathway. Hence, if you wish to produce hydrogen through biological means, it becomes necessary to intervene in the anaerobic digestion process eliminating as many hydrogen consuming microorganisms as possible [12]. The required microbial population control can initially be achieved by treatment of the inocula [13] and later, during fermentation, through the control of process parameters.

Table 1. Modified from [6]

Equations	ΔG^0 (kJ/reaction)
Proton-reducing (H_2-producing) acetogenic bacteria	
A $CH_3CH_2CH_2COO^- + 2H_2O \rightleftharpoons 2CH_3COO^- + 2H_2 + H^+$	+48.1
B $CH_3CH_2COO^- + 3H_2O \rightleftharpoons CH_3COO^- + HCO_3^- + H^+ + 3H_2$	+76.1
H_2-using methanogens:	
C $4H_2 + HCO_3^- + H+ \rightleftharpoons CH_4 + 3H_2O$	-135.6
Co-culture of 1 and 2	
A+C $2CH_3CH_2CH_2COO^- + HCO_3^- + H_2O \rightleftharpoons 4CH_3COO^- + H^+ + CH_4$	-39.4
B+C $4CH_3CH_2COO^- + 12H_2 \rightleftharpoons 4CH_3COO^- + HCO_3^- + H^+ + 3CH_4$	-102.4

2.2. Effects of Microbial Population Control

It is important to observe the close relationship that exists between the production of hydrogen, acetogenesis and methanogenesis. In AD, most of the hydrogen will be generated during acetogenesis [6]. However, due to thermodynamic restrictions imposed by Gibbs standard energy (ΔG^0), production of hydrogen in this stage will become inhibited if hydrogen partial pressure (pH_2) increases over a threshold level, which is defined by the particular hydrogen production reaction taking place. In pure cultures of H_2-producing acetogenic bacteria (Table 1 - modified from [6]), the oxidation of propionate and butyrate to acetate and hydrogen (equations A and B), will be prevented by thermodynamic restrictions. In this case the ΔG^0 values for the corresponding oxidation reactions are positive and therefore the reactions will not proceed. Conversely, if H_2-producing acetogenic bacteria are co-cultured with methanogenic bacteria (equation C), that will consume hydrogen as soon as it is produced, the ΔG^0 values become negative and the reactions (equations A+C and B+C) can proceed towards the formation of acetate and methane.

To summarize, biological hydrogen can be produced during acidogenic fermentation and during acetogenesis. Microbial population control, which eliminates as many H_2-consuming microorganisms as possible, prevents the production of hydrogen through acetogenesis due to the loss of H_2-consuming syntrophic partners. In the absence of methanogenic or other H_2 consuming bacteria, hydrogen partial pressure very quickly reaches values that inhibit hydrogen production by acetogenic bacteria. For example, from the data provided by Goen Ho et. al. [14], it can be calculated that very low pH_2 values will inhibit the oxidation of propionate (3.279 Pa) and butyrate (5.698 Pa) when the following conditions are considered, 1M is the concentration for all species, pH=7, T=298°K, pressure = 1 atm and $\Delta G = 0$. Note that ΔG is not Gibbs standard energy (ΔG^0). ΔG is the Gibbs energy value corrected for biological systems [14] using the equation $\Delta G = \Delta G^0 + RT.ln([products]/[substrates])$ where R = universal gas constant and T = absolute temperature.

In opposition, acidogenic fermentation of glucose does not present a thermodynamic limitation to hydrogen production. Again, the information provided by Goen Ho et. al. shows that the reaction of conversion of glucose to hydrogen (equation 8) will only be inhibited under extremely high pressures (1.1633 x 10^{14} Pa), that are not expected to occur in biological systems. Similar results were found by Ginkel et. al. [15], whom under the conditions established by their experimental method determined that, in theory, the pH_2 could reach values close to 7.49 x 10^{14} pa during glucose fermentation. It must be stated that this does not mean that acidogenic glucose fermentation is not influenced by pH_2, as it is proven by a large number of scientific publications [12, 14, 16, 17], lower hydrogen partial pressures will result in higher hydrogen yields. The reason for this is that the enzymes involved in the production of hydrogen through glucose fermentation can be thermodynamically regulated by hydrogen concentration [17]. In mixed cultures as pH_2 increases, the fermentation of sugars derived from carbohydrates becomes the primary (if not the only) source of hydrogen.

$$C_6H_{12}O_6 + 4H_2O \longrightarrow 2CH_3COOH + 2HCO_3^- + 4H_2 + 4H^+ \tag{8}$$

Acidogenic fermentation of glucose [14]. $\Delta G^0 = -206,1\ kJ/reaction$

2.3. Fermentation of Carbohydrate Derived Sugars

Under anaerobic conditions, once complex carbohydrates have been broken down to simple sugars by the action of hydrolytic bacteria, soluble monomers (e.g. glucose) can be taken up by fermentative bacteria and degraded to pyruvate through the Embden-Meyerhof (EM) pathway [18]. Two molecules of adenosine triphosphate (ATP) are produced for each molecule of glucose that enters the EM pathway and therefore energy is gained. Pyruvate can then be converted to a number of different fermentation end-products such as formate, acetate, lactate, propionate, succinate, butyrate, ethanol, propanol, butanol and acetone [18]. Hydrogen yield is directly related to the type of metabolite formed and higher yields are associated with the formation of acetate [12, 14]. As explained by R. Stegmann [9], theoretically 4 mol of hydrogen per mol of glucose could be obtained with acetate as the fermentation end product (equation 9) whereas if butyrate or propionate is produced only 2 moles (equation 10) and 1 mole of hydrogen per mol of glucose will be obtained, respectively. For mixed cultures, fermentation usually produces a mixture of acetic and butyric acid (equation 2) and a theoretical maximum of 2 mol of hydrogen can be attained per mol of glucose. In addition, fermentation generates other reduced products such as lactate and alcohols, these metabolites retain hydrogen atoms that cannot be used to produce hydrogen gas. Therefore in order to achieve higher yields microbial metabolism has to be directed towards the production of volatile fatty acids [14, 16], such as acetate, butyrate and propionate.

$$C_6H_{12}O_6 + 2H_2O \longrightarrow 2CH_3COOH + 2CO_2 + 4H_2 \tag{9}$$

Glucose fermented to acetic acid.

$$C_6H_{12}O_6 \longrightarrow CH_3(CH_2)_2COOH + 2CO_2 + 2H_2 \tag{10}$$

Glucose fermented butyric acid.

The metabolism of *Clostridium* spp. (Figure 2, modified from [17]) has been extensively studied since some *Clostridium* spp. are strong hydrogen producers [16] and have been found to be the most common species in bioreactors designed for hydrogen production by several authors [17]. The enzymes pyruvate:ferredoxin oxidoreductase (PFOR) and

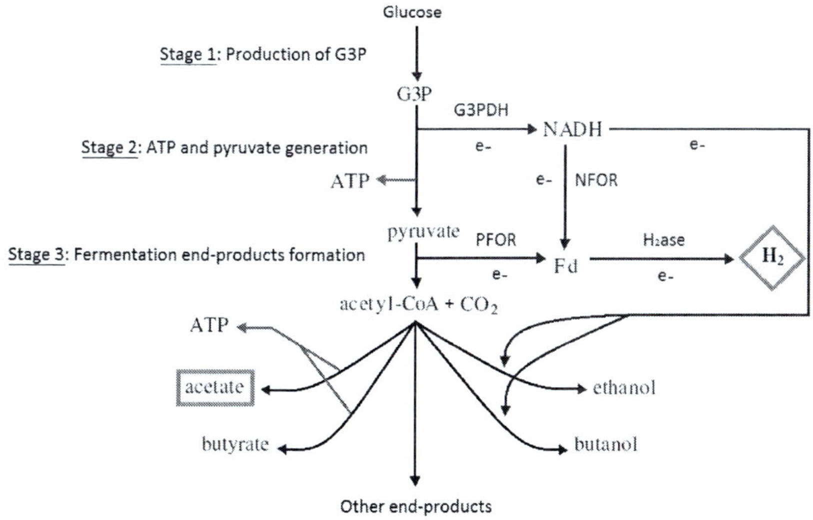

Figure 2. Modified from 17: Emden-Meyerhof pathways for glucose fermentation in *Clostridium* spp. Black lines indicate substrate transformation and blue lines represent electron flow. G3P = glyceraldehyde-3-phosphate; G3PDH = glyceraldehyde-3-phosphate dehydrogenase;PFOR = pyruvate:ferredoxin oxidoreductase; NFOR = nicotinamide-adenine dinucleotide:ferredoxin oxidoreductase; Fd = ferredoxin; H_2ase = hydrogenase; NADH = nicotinamide-adenine dinucleotide; ATP = adenosine triphosphate. Acetate is the preferred fermentation end-product from the stand point of hydrogen production.

nicotinamide-adenine dinucleotide:ferredoxin oxidoreductase (NFOR) are of key importance in biohydrogen production since they will reduce the molecule ferredoxin (Fd) from which, in turn, H_2 will be produced by the action of another important enzyme, the hydrogenase (H_2ase). The EM pathway can be divided into three stages, in the first one, that does not involve oxidation-reduction reactions, glyceraldehyde-3-phosphate (G3P) is produced from glucose. Then, in the second stage in which a series of oxidation-reduction reactions occur, as G3P is oxidized to pyruvate electrons are transferred to nicotinamide-adenine dinucleotide (NADH). These electrons can then be relocated by NFOR to ferredoxin contributing to the accumulation of reduced Fd. In the third stage, electrons are extracted from pyruvate and transferred to Fd as pyruvate is oxidized by PFOR to acetyl-CoA (equation 11) and carbon dioxide. Biological hydrogen is produced by re-oxidation of Fd (equation 12) coupled to proton reduction through the action of the H_2ase enzyme.

$$\text{Pyruvate} + \text{CoA} + 2\,\text{Fd(oxidized)} \longrightarrow \text{Acetyl-CoA} + 2\,\text{Fd(reduced)} + CO_2 \quad (11)$$

Pyruvate oxidation to Acetyl-CoA [16].

$$2\,\text{Fd(reduced)} \longrightarrow 2\,\text{Fd(oxidized)} + H_2 \quad (12)$$

Fd oxidation by hydrogenase enzyme [16].

The final destination of acetyl-CoA depends on the microbial population involved and on other factors such as hydrogen partial pressure. NFOR is sensitive to pH_2 and it will be inhibited by pressures higher than 10 Pa [14], for that reason less reduced Fd will be available for H_2 production. Additionally NADH has to be re-oxidized so that bacterial cells can maintain redox balance, therefore electrons from NADH are used to generate more reduced end-products [17], such as ethanol and butanol and propionate. Higher pH_2 will result in lower H_2 yields and in a metabolic shift away from acetate.

3. Final Considerations Regarding the Production of Hydrogen with Mixed Cultures

The objective of the precedent sections was to detail the metabolic pathways through which hydrogen is produced and therefore emphasis was made on the effects of hydrogen partial pressure since it can directly influence the pathways preferred by bacteria to produce hydrogen. However, if high yields of hydrogen are to be attained, many other process parameters such as substrate type, nutrients composition, pH, organic load rate and temperature are to be carefully considered and optimized for hydrogen production in accordance with the particular mixed culture chosen to perform the fermentation.

References

[1] U.S. Environmental Protection Agency. Renewable and alternative fuels - fuels and fuel additives. http://www.epa.gov/otaq/fuels/alternative-renewablefuels/ index.htm, 18th July 2012.

[2] L. Gómez J.L.G. Fierro and M.A. Pe na. El hidrogeno: Un vector energetico no contaminante para automocion. *Elsevier Science Publishers*, 110, 1997.

[3] Hydrogen safety. \smallhttp://www.cab.cnea.gov.ar/ieds/extras/hojitas_conocimiento/seguridad/pag_11_y_12_seguridad_del_hidrogeno-aprea.pdf, 2009.

[4] I. Ntaikou, G. Antonopoulou, and G. Lyberatos. Biohydrogen production from biomass and wastes via dark fermentation: A review. *Waste and Biomass Valorization*, 1(1):21, 2010.

[5] Verónica L. Martínez, Rodrigo E. García, Gustavo Curutchet, Alfredo Sanguinetti, Héctor J. Fasoli, and Juan I. Franco. Demonstration of the possibility to power a fuel cell with hydrogen derived from the fermentation of sugar. *International Journal of Hydrogen Energy*, (0):–, 2012.

[6] FAO Agriculture and Consumer Protection Department. http://www.fao.org/docrep/w7241e/w7241e0f.htm, year = 1997,.

[7] Karena Ostrem. Greening waste: Anaerobic digestion for treating the organic fraction of municipal solid waste. Master's thesis, Columbia University, 2004.

[8] Lise Appels, Jan Baeyens, Jan Degrève, and Raf Dewil. Principles and potential of the anaerobic digestion of waste-activated sludge. *Progress in Energy and Combustion Science*, 34(6):755 – 781, 2008.

[9] M. MEYER D. ZURAWSKI and R. STEGMANN. Fermentative production of biohydrogen from biowaste using digested sewage sludge as inoculum. *Tenth International Waste Management and Landfill Symposium*, 2005.

[10] Dra. Cristina Cavinato. Anaerobic digestion fundamentals I. `http://www.valorgas.soton.ac.uk/Pub_docs/JyU%20SS%202011/CC%201.pdf`, August 2011.

[11] Volker Müller. Energy conservation in acetogenic bacteria. *Applied and Environmental Microbiology*, 69(11):6345 – 6353, 2003.

[12] Carlos Dinamarca and Rune Bakke. Process parameters affecting the sustainability of fermentative hydrogen production: A short-review. *International Journal of Energy and Environment*, 2(6):1067 – 1078, 2011.

[13] Chaganti Subba Rao, Kim Dong-Hoon, and Lalman Jerald A. Dark fermentative hydrogen production by mixed anaerobic cultures: Effect of inoculum treatment methods on hydrogen yield. *Renewable Energy*, n.d.

[14] Ka Yu Cheng, Ralf Cord-Ruwisch, and Goen Ho. Limitations of bio-hydrogen production by anaerobic fermentation process: An overview. *AIP Conference Proceedings*, 941(1):264 – 269, 2007.

[15] Bruce E Logan, Sang-Eun Oh, In S Kim, and Steven Van Ginkel. Biological hydrogen production measured in batch anaerobic respirometers. *Environmental Science and Technology*, 36(11):2530 – 2535, 2002.

[16] Kaushik Nath and Debabrata Das. Improvement of fermentative hydrogen production: various approaches. *Applied Microbiology and Biotechnology*, 65(5):520 – 529, 2004.

[17] Jeremy T. Kraemer and David M. Bagley. Improving the yield from fermentative hydrogen production. *Biotechnology Letters*, 29(5):685 – 695, 2007.

[18] Gerard J. Tortora, Berdell R. Funke, and Christine L. Case. *Microbiology: An Introduction*. Pearson Benjamin Cummings, 2010.

INDEX

A

B

C

D

E

F

G

H

O

P

Q

R

S

T

U

V

W

Y

Z